Vector-Valued Functions and their Applications

Mathematics and Its Applications (*Chinese Series*)

Volume 3

Vector-Valued Functions and their Applications

by

Chuang-Gan Hu

Department of Mathematics,
Nankai University,
Tianjin, People's Republic of China

and

Chung-Chun Yang

Department of Mathematics,
The Hong Kong University of Science and Technology,
Kowloon, Hong Kong

KLUWER ACADEMIC PUBLISHERS

DORDRECHT / BOSTON / LONDON

Library of Congress Cataloging-in-Publication Data

Hu, Chuang-gan, 1935-
Vector-valued functions and their applications / by Chuang-Gan Hu and Chung-Chun Yang.
p. cm. -- (Mathematics and its applications. Chinese series ; 3)
Includes bibliographical references (p.) and index.
ISBN 0-7923-1605-3 (HB : acid free paper)
1. Vector valued functions. I. Yang, Chung-Chun, 1942- II. Title. III. Series: Mathematics and its applications. Chinese series ; 3.
QA331.7.H82 1992
515'.9--dc20 91-46130

ISBN 0-7923-1605-3

Published by Kluwer Academic Publishers,
P.O. Box 17, 3300 AA Dordrecht, The Netherlands.

Kluwer Academic Publishers incorporates
the publishing programmes of
D. Reidel, Martinus Nijhoff, Dr W. Junk and MTP Press.

Sold and distributed in the U.S.A. and Canada
by Kluwer Academic Publishers,
101 Philip Drive, Norwell, MA 02061, U.S.A.

In all other countries, sold and distributed
by Kluwer Academic Publishers Group,
P.O. Box 322, 3300 AH Dordrecht, The Netherlands.

Printed on acid-free paper

Printed in the Netherlands

SERIES EDITOR'S PREFACE

'Et moi, ..., si j'avait su comment en revenir, je n'y serais point allé.'

Jules Verne

The series is divergent; therefore we may be able to do something with it.

O. Heaviside

One service mathematics has rendered the human race. It has put common sense back where it belongs, on the topmost shelf next to the dusty canister labelled 'discarded nonsense'.

Eric T. Bell

Mathematics is a tool for thought. A highly necessary tool in a world where both feedback and nonlinearities abound. Similarly, all kinds of parts of mathematics serve as tools for other parts and for other sciences.

Applying a simple rewriting rule to the quote on the right above one finds such statements as: 'One service topology has rendered mathematical physics ...'; 'One service logic has rendered computer science ...'; 'One service category theory has rendered mathematics ...'. All arguably true. And all statements obtainable this way form part of the raison d'être of this series.

This series, *Mathematics and Its Applications*, started in 1977. Now that over one hundred volumes have appeared it seems opportune to reexamine its scope. At the time I wrote

> "Growing specialization and diversification have brought a host of monographs and textbooks on increasingly specialized topics. However, the 'tree' of knowledge of mathematics and related fields does not grow only by putting forth new branches. It also happens, quite often in fact, that branches which were thought to be completely disparate are suddenly seen to be related. Further, the kind and level of sophistication of mathematics applied in various sciences has changed drastically in recent years: measure theory is used (non-trivially) in regional and theoretical economics; algebraic geometry interacts with physics; the Minkowsky lemma, coding theory and the structure of water meet one another in packing and covering theory; quantum fields, crystal defects and mathematical programming profit from homotopy theory; Lie algebras are relevant to filtering; and prediction and electrical engineering can use Stein spaces. And in addition to this there are such new emerging subdisciplines as 'experimental mathematics', 'CFD', 'completely integrable systems', 'chaos, synergetics and large-scale order', which are almost impossible to fit into the existing classification schemes. They draw upon widely different sections of mathematics."

By and large, all this still applies today. It is still true that at first sight mathematics seems rather fragmented and that to find, see, and exploit the deeper underlying interrelations more effort is needed and so are books that can help mathematicians and scientists do so. Accordingly MIA will continue to try to make such books available.

If anything, the description I gave in 1977 is now an understatement. To the examples of interaction areas one should add string theory where Riemann surfaces, algebraic geometry, modular functions, knots, quantum field theory, Kac-Moody algebras, monstrous moonshine (and more) all come together. And to the examples of things which can be usefully applied let me add the topic 'finite geometry'; a combination of words which sounds like it might not even exist, let alone be applicable. And yet it is being applied: to statistics via designs, to radar/sonar detection arrays (via finite projective planes), and to bus connections of VLSI chips (via difference sets). There seems to be no part of (so-called pure) mathematics that is not in immediate danger of being applied. And, accordingly, the applied mathematician needs to be aware of much more. Besides analysis and numerics, the traditional workhorses, he may need all kinds of combinatorics, algebra, probability, and so on.

In addition, the applied scientist needs to cope increasingly with the nonlinear world and the extra mathematical sophistication that this requires. For that is where the rewards are. Linear models are honest and a bit sad and depressing: proportional efforts and results. It is in the nonlinear world that infinitesimal inputs may result in macroscopic outputs (or vice versa). To appreciate what I am hinting at: if electronics were linear we would have no fun with transistors and computers; we would have no TV; in fact you would not be reading these lines.

There is also no safety in ignoring such outlandish things as nonstandard analysis, superspace and anticommuting integration, p-adic and ultrametric space. All three have applications in both electrical engineering and physics. Once, complex numbers were equally outlandish, but they frequently proved the shortest path between 'real' results. Similarly, the first two topics named have already provided a number of 'wormhole' paths. There is no telling where all this is leading - fortunately.

Thus the original scope of the series, which for various (sound) reasons now comprises five subseries: white (Japan), yellow (China), red (USSR), blue (Eastern Europe), and green (everything else), still applies. It has been enlarged a bit to include books treating of the tools from one subdiscipline which are used in others. Thus the series still aims at books dealing with:

- a central concept which plays an important role in several different mathematical and/or scientific specialization areas;
- new applications of the results and ideas from one area of scientific endeavour into another;
- influences which the results, problems and concepts of one field of enquiry have, and have had, on the development of another.

As the authors write in their preface, vector-valued functions turn up everywhere and they are a fundamental tool in physics, spectral theory, approximation and many other fields. Yet there was, so far, no systematic treatise on the topic. Here is one by two authors who have done quite a bit of work in the field.

At first sight one could think that a vector-valued function (instead of a scalar-valued one) would not bring so much new in the way of problems and theory. Just look at the components. It is far otherwise. There is a great deal more to vectorspaces, finite and infinite dimensional, then to scalars, and this takes an added emphasis in the case of vector-valued functions. To try to explain how much would take me far beyond the confines of an editorial preface; even if I could. It seems better to leave that to the authors, and the reader who wants to find out is therefore encouraged to peruse this unique volume.

The shortest path between two truths in the real domain passes through the complex domain.

J. Hadamard

La physique ne nous donne pas seulement l'occasion de résoudre des problèmes ... elle nous fait pressentir la solution.

H. Poincaré

Never lend books, for no one ever returns them; the only books I have in my library are books that other folk have lent me.

Anatole France

The function of an expert is not to be more right than other people, but to be wrong for more sophisticated reasons.

David Butler

Bussum, 9 February 1992

Michiel Hazewinkel

TABLE OF CONTENTS

PREFACE

The theory of vector-valued functions with one variable is one of the fundamental tools in studying modern physics, the spectral theory of operators, approximation of analytic operators, analytic mappings between vectors, and vector-valued functions of several variables. It seems that, thus far, no book specialized in dealing with vector-valued functions of one variable has been published in the West, except for some brief introduction to such functions in some books dealing with functional analysis or function theory. The present book will be a remedy for satisfying such a need. It consists of three chapters: theory of normal functions, H^p space, and vector-valued functions with their applications, and it incorporates a lot of original research work obtained by the authors and others. The reader of this book is assumed to have a basic knowledge in real and complex analysis. The material of this book has been used as lecture notes for an optional course for senior mathematics and physics students at Nankai University. The approach of combining together function theory, functional analysis, and the theory of boundary value problems and integral equations has not only made this book unique, but has also benefited students in developing their ability in abstracting and generalizing mathematical theories.

We would like to express our sincere thanks to Dr. D.J. Larner, the publisher of Kluwer Academic Publishers, for his interest and support in publishing this book. We'd like also to acknowledge the excellent typing job done by the Apparatus Office of Nankai University and Tianli Company, and to thank our students Mr. Z.J. Wang, Ms. D.P. Zhang and Mr. G.T. Ding for checking the final proof of this book.

C.G. Hu
C.C. Yang

CHAPTER 1

Theory of Normal Families

1. Preliminaries

1.1. THE DEFINITIONS OF $n(r,a)$, $N(r,a)$, $b(r,f)$, $q(r,f)$, $m(r,f)$, $m_2(r,f)$, $A(r,f)$, $S(r,f)$, AND $T(r,f)$

Let $f(z)$ be a function meromorphic in the disc $|z|<R(R<\infty)$. To any r satisfying $0\leqslant r<R$, we define $n(r,f,a)$, or simply, without confusion, $n(r,a)$ to be the number of the roots of the equation: $f(z)=a$ in $|z|<r$; a root of multiplicity k will be counted k times.

We set

$$N(r,a)=\int_0^r \frac{n(t,a)-n(0,a)}{t}dt+n(0,a)\log r.$$

In order to avoid confusion, when $a\neq\infty$, we use $n(r,1/(f-a))$ and $N(r,1/(f-a))$ instead of $n(r,a)$ and $N(r,a)$ respectively, and when $a=\infty$, the notations $n(r,f)$ and $N(r,f)$ are used for $n(r,\infty)$ and $N(r,\infty)$ respectively.

Set

$$b(r,f)=\frac{1}{4\pi}\int_0^{2\pi}\frac{\partial \log(1+|f(re^{i\theta})|^2)}{\partial r}rd\theta.$$

$$q(r,f)=\frac{1}{4\pi}\int_0^{2\pi}\log(1+|f(re^{i\theta})|^2)d\theta.$$

$$m(r,f)=\frac{1}{2\pi}\int_0^{2\pi} log^+(1+|f(re^{i\theta})|)d\theta,$$

$$(log^+x=max(log\ x,0)).$$

$$m_2(r,f)=\frac{1}{2\pi}\int_0^{2\pi}(log^+|f(re^{i\theta})|)^2d\theta.$$

$$A(r,f)=\frac{1}{\pi}\int_0^r\int_0^{2\pi}\frac{|f'(te^{i\theta})|^2}{(1+|f(te^{i\theta})|^2)^2}t\,d\theta\,dt.$$

$$S(r,f)=\int_0^r\frac{A(t,f)}{t}dt.$$

$$T(r,f)=m(r,f)+N(r,f).$$

The properties and interrelationships among these functions will be discussed in the following.

1.2. Convex function

Let $\varphi(r)$ be a real-valued continuous function defined in the interval $[R_1,R_2]$ and suppose that to any $r_j (j=1,2,3)$ satisfying $R_1 \leqslant r_1 < r_2 < r_3 < R_3$ the inequality:

$$\varphi(r_3)\frac{r_2-r_1}{r_3-r_1}+\varphi(r_1)\frac{r_3-r_2}{r_3-r_1}\geqslant\varphi(r_2).$$

holds, then $\varphi(r)$ is called a convex function on $[R_1,R_2]$.

Theorem 1.1.1. *Let $\varphi(r)$ be a convex function on (R_1,R_2). Then to any positive numbers t_j and points x_j, $j=1,\ldots n$, in (R_1,R_2).*

$$\varphi\left(\frac{\sum_{j=1}^n t_jx_j}{\sum_{j=1}^n t_j}\right)\leqslant\frac{\sum_{j=1}^n t_j\varphi(x_j)}{\sum_{j=1}^n t_j}.$$

Proof. Let $r_1=x_1$, $r_2=\frac{t_1x_1+t_2x_2}{t_1+t_2}$, $r_3=x_2$. Then by the definition of a convex function, we have

$$\varphi\left(\frac{t_1x_1+t_2x_2}{t_1+t_2}\right)\leqslant\frac{t_1\varphi(x_1)+t_2\varphi(x_2)}{t_1+t_2}.$$

The theorem follows by standard mathematical induction argument.

Theorem 1.1.2. *Let $\mu(r)$ and $g(r)$ be two continuous functions definded on the closed interval*

$[r_1, r_2]$ *with* $T_1 \leqslant g(r) \leqslant T_2$ *and* $\int_{r_1}^{r_2} \mu(r)dr > 0$. *Assume that* $\varphi(t)$ *is a continuous convex function definded on* $[T_1, T_2]$, *then*

$$\varphi\left[\frac{\int_{r_1}^{r_2} \mu(r)g(r)dr}{\int_{r_1}^{r_2} \mu(r)dr}\right] \leqslant \frac{\int_{r_1}^{r_2} \mu(r)\varphi(g(r))dr}{\int_{r_1}^{r_2} \mu(r)dr}.$$

Proof. Let

$$\mu_{v_n} = \mu\ (r_1 + v\frac{r_2 - r_1}{n}).$$

$$g_{v_n} = g(r_1 + v\frac{r_2 - r_1}{n})\ (v = 1, 2, \cdots, n).$$

Then, according to Theorem1. 1. 1, we have

$$\varphi(\frac{\mu_{1n}g_{1n} + \cdots \mu_{nn}g_{nn}}{\mu_{1n} + \cdots + \mu_{nn}}) \leqslant \frac{\mu_{1n}\varphi(g_{1n}) + \cdots + \mu_{nn}\varphi(g_{nn})}{\mu_{1n} + \cdots + \mu_{nn}}.$$

By the very definition of an integral, it follows that when $n \to \infty$,

$$\frac{r_2 - r_1}{n}(\sum_{j=1}^{n} \mu_{jn}g_{jn}) \to \int_{r_1}^{r_2} \mu(r)g(r)dr,$$

$$\frac{r_2 - r_1}{n}(\sum_{j=1}^{n} \mu_{jn}) \to \int_{r_1}^{r_2} \mu(r)dr,$$

and

$$\frac{r_2 - r_1}{n}[\sum_{j=1}^{n} \mu_{jn}\varphi(g_{jn})] \to \int_{r_1}^{r_2} \mu(r)\varphi(g(r))dr.$$

The conclusion follows.

Corollary. $\log[\frac{1}{r_2 - r_1}\int_{r_1}^{r_2} g(r)dr] \geqslant \frac{1}{r_2 - r_1}\int_{r_1}^{r_2} \log g(r)dr.$

Proof. The assertion follows immediately by virtue of Theorem 1. 1. 1 by letting $\varphi(t) = -\log t$, $\mu(r) = 1$, and $g(r) > 0$.

1. 3. THE PROPERTIES OF $A(r, f)$

Lemma 1. 1. 1. *Let* $f(z)$ *be a regular function in domain* D, *then*

$$\triangle |f(z)|^2 = 4|f'(z)|^2;$$

$$\triangle \log(1+|f(z)|^2)=\frac{4|f'(z)|^2}{(1+|f(z)|^2)^2}$$

where the operator $\triangle u$ *represents the Laplacian*: $\frac{\partial^2 u}{\partial x^2}+\frac{\partial^2 u}{\partial y^2}$.

Proof. *Let* $f(z)=u(x,y)+v(x,y)i$, where u,v are harmonic functions; hence $\triangle u=0$, $\triangle v=0$.

By a simple calculation, we have

$$\begin{aligned}&\triangle|f(z)|^2\\&=2[(\frac{\partial u}{\partial x})^2+(\frac{\partial v}{\partial x})^2+(\frac{\partial u}{\partial y})^2+(\frac{\partial v}{\partial y})^2+u\triangle u+v\triangle v]\\&=4|f'(z)|^2\\&\triangle \log(1+|f(z)|^2)\\&=\frac{\triangle|f(z)|^2+|f|^2\triangle|f|^2-[(\frac{\partial}{\partial x}|f|^2)^2+(\frac{\partial}{\partial y}|f|^2)^2]}{(1+|f|^2)^2}\end{aligned}$$

But

$$\begin{aligned}&(\frac{\partial}{\partial x}|f|^2)^2+(\frac{\partial}{\partial y}|f|^2)^2\\&=4(u^2+v^2)[(\frac{\partial u}{\partial x})^2+(\frac{\partial y}{\partial x})^2]\\&=4|f|^2|f'|^2.\end{aligned}$$

Hence

$$\triangle \log(1+|f|^2)=\frac{4|f'|^2}{(1+|f|^2)^2}$$

Theorem 1.1.3. $A(r,f)=b(r,f)+n(r,f)$.

Proof. First of all, $f(z)$ has only finitely number of poles in the disk: $|z|\leqslant r\,(r<\infty)$. Also let $C_1,C_2,\cdots,C_k$ be disks centered at these poles with sufficiently small radii all lying in: $|z|\leqslant r$ and separated from each other.

Let D_0 be the region of $|z|<r$ after removing those disks mentioned above. Then by the lemma we have

$$\iint_{D_0}\triangle \log(1+|f(te^{i\theta})|^2)d\Omega=\iint_{D_0}\frac{4|f'(te^{i\theta})|^2}{(1+|f(te^{i\theta})|^2)^2}d\Omega, \tag{1.1.1}$$

where $d\Omega$ denotes the surface element of D_0. It follows from Green's theorem that

$$\iint_{D_0} \triangle \log(1+|f(te^{i\theta})|^2)\, d\Omega$$

$$= \int_0^{2\pi} \frac{\partial \log(1+|f(re^{i\theta})|^2)}{\partial r} r\, d\theta - \sum_{|b_\nu|\leqslant r} \int_0^{2\pi} \frac{\partial \log(1+|f(z_\nu)|^2)}{\partial \rho_\nu} \rho_\nu\, d\theta_\nu,$$

where $\rho_\nu e^{i\theta_\nu} = z_\nu - b_\nu$, z lies on the circumference of C_ν and $\sum_{|b_\nu|\leqslant r}$ means the sum of the integrations along all the circles C_ν centered at b_ν that lie in $|z|\leqslant r$.

Letting $\rho_\nu \to 0$, the limit of the integration on the right hand side of (1. 1. 1) tends to $4\pi A(r, f)$.

Let $f(z)$ be expressed as $\frac{h_1(z)}{h_2(z)}$, where $h_j(z)$ $(j=1,2)$ are regular in the disk: $|z|<r$. Then

$$\int_0^{2\pi} \frac{\partial \log(1+|f(z_\nu)|^2)}{\partial \rho_\nu} \rho_\nu\, d\theta_\nu$$

$$= \int_0^{2\pi} \frac{\partial \log(|h_1(x_\nu)|^2+|h_2(z_\nu)|^2)}{\partial \rho_\nu} \rho_\nu\, d\theta$$

$$-2 \int_0^{2\pi} \frac{\partial \log|h_2(z_\nu)|}{\partial \rho_\nu} \rho_\nu\, d\theta.$$

Clearly, when $\rho_\nu \to 0$, the first term of the right side of the above equation tends to zero. While the second term on the right side of the equation is equal to

$$-2 \int_0^{2\pi} \frac{\partial \log|h_2(z_\nu)|}{\partial \rho_\nu} \rho_\nu\, d\theta$$

$$= -4\pi \cdot \frac{1}{2\pi} \int_0^{2\pi} \frac{\partial \log|h_2(z_\nu)|}{\partial \log \rho_\nu} d\theta_\nu = -4\pi n_\nu,$$

where n_ν is the order of the pole b_ν. Therefore, when $\rho_\nu \to 0$,

$$\iint_{D_0} \triangle \log(1+|f(te^{i\theta})|^2)\, d\Omega \to$$

$$\int_0^{2\pi} \frac{\partial \log(1+|f(re^{i\theta})|^2)}{\partial r} r\, d\theta + \sum_{|b_\nu|\leqslant r} 4\pi n_\nu.$$

It follows that

$$A(r, f) = b(r, f) + n(r, \infty).$$

1. 4. THE PROPERTIES OF $S(r, f)$.

Theorem 1. 1. 4. *Let $f(z)$ be a function meromorphic in $|z|\leqslant r$, which is regular at point $z=0$.*

Then

$$S(r,f)=g(r,f)+N(r,\infty)-\frac{1}{2}log(1+|f(0)|).$$

Proof. Clearly $\frac{\partial\, log(1+|f(te^{i\theta})|^2)}{\partial t}$ has only a finite number of poles on $|z|=t(0<t<r)$. Let the moduli of these poles be denoted by $t_j(j=1,2,\cdots,k)$ and assume

$$0<t_1<t_2<\cdots<t_k<r.$$

By choosing sufficiently small ε' $(<\frac{r_{j+1}-t_j}{2})$ we have

$$\int_{t_j+\varepsilon}^{t_{j+1}-\varepsilon'}\frac{A(t,f)}{t}dt$$

$$=\int_{t_j+\varepsilon'}^{t_{j+1}-\varepsilon'}\left(\frac{1}{4\pi}\int_0^{2\pi}\frac{\partial\, log(1+|f|)^2}{\partial t}t\, d\theta\right)\frac{dt}{t}+\int_{t_j+\varepsilon'}^{t_{j+1}-\varepsilon'}\frac{n(t,\infty)}{t}dt$$

$$=\frac{1}{4\pi}\int_0^{2\pi}\left[log(1+|f(te^{i\theta})|^2)\right]_{t=t_j+\varepsilon'}^{t=t_{j+1}-\varepsilon'}d\theta+\int_{t_j+\varepsilon'}^{t_{j+1}-\varepsilon'}\frac{n(t,\infty)}{t}dt$$

$$=q(t_{j+1}-\varepsilon',f)-q(t_j+\varepsilon',f)+\int_{t_j+\varepsilon'}^{t_{j+1}-\varepsilon'}\frac{n(t,\infty)}{t}dt.$$

Now letting $\varepsilon'\to 0$ and by the continuity of $q(r,f)$, we obtain

$$\int_{t_j}^{t_{j+1}}\frac{A(t,f)}{t}dt=q(t_{j+1},f)-q(t_j,f)$$

$$+\int_{t_j}^{t_{j+1}}\frac{n(t,\infty)}{t}dt\quad(j=1,2,\cdots,k-1).$$

Clearly similar relations hold in $(\varepsilon,t_1),\cdots,(t_k,r)$; $0<\varepsilon<t$.

Hence

$$\int_\varepsilon^r\frac{A(t,f)}{t}dt=q(r,f)-q(\varepsilon,f)+\int_\varepsilon^r\frac{n(t,\infty)}{t}dt$$

$$=q(r,f)+\int_\varepsilon^r\frac{n(t,\infty)-n(0,\infty)}{t}dt$$

$$+n(0,\infty)log\ r-[q(\varepsilon,f)+n(0,\infty)log\ \varepsilon].$$

Now by assumption that $f(z)$ is regular at $z=0$, it follows that $n(0,\infty)=0$, and hence

$$\lim_{\varepsilon\to 0}q(\varepsilon,f)=\frac{1}{2}log(1+|f(0)|^2).$$

Theorem 1.1.5. *$S(r,f)$ is a positive and monotonic increasing continuous function of r. Moreover, it is also a convex function of $log\ r$.*

Proof. The first part is obvious. As to the second part, the assertion follows by

$$\frac{dS(r,f)}{d\log r}=A(r,f).$$

and the fact that $A(r,f)$ is a monotonic increasing function of r.

Theorem 1. 1. 6. *$N(r,w)$ is a monotonic increasing convex function of $\log r$. Moreover*

$$N(r,w)-n(0,w)\log r=\sum_{0<|a_\nu(w)|\leqslant r}\log\frac{r}{|a_\nu(w)|},$$

where $a_\nu(w)$ denotes a point of w-point of f with absolute value no greater than r.

Proof. By considering $\dfrac{dN(r,w)}{d\log r}$ the first assertion follows.

Let r_j be the absolute values of the w-points of $f(z)$ with $0<r_1<r_2\cdots<r_n<r$. Let n_j denote the number of w-points of f on $|z|=r_j$. Then

$$\int_0^r\frac{n(t,w)-n(0,w)}{t}dt$$

$$=\sum_{j=1}^{n}\int_{r_j}^{r_{j+1}}\frac{n(t,w)-n(0,w)}{t}dt$$

$$=\sum_{j=1}^{n}\log\left(\frac{r_{j+1}}{r_j}\right)^{n_1+\cdots+n_j}$$

$$=\sum_{0<|a_\nu(w)|\leqslant r}\log\frac{r}{|a_\nu(w)|}.$$

The theorem is thus proved.

Theorem 1. 1. 7. *Let*

$$C(f)=\begin{cases}-\dfrac{1}{2}\log(1+|f(0)|^2) & (\text{when } f(0)\neq\infty)\\ -\log|C_{-m}| & (\text{when } f(0)=\infty)\end{cases}$$

Then

$$C(f)\leqslant S(r,f)-T(r,f)\leqslant\frac{\log 2}{2}+C(f).$$

Proof. Applying Theorem 1. 1. 4 and by the definition of $T(r,f)$, we obtain

$$S(r,f)-T(r,f)=q(r,f)-m(r,f)+C(f).$$

But

$$\frac{1}{4\pi}\int_0^{2\pi} log(1+|f(re^{i\theta})|^2)d\theta$$
$$=\frac{1}{4\pi}\int_0^{2\pi} log^+(1+|f(re^{i\theta})|^2)\, d\theta$$
$$\leqslant\frac{1}{4\pi}\int_0^{2\pi} log^+|(fre^{i\theta})|^2 d\theta+\frac{1}{4\pi}\int_0^{2\pi} log2\, d\theta$$
$$=m(r,\infty)+\frac{log2}{2}.$$

On the other hand,

$$\frac{1}{4\pi}\int_0^{2\pi} log^+(1+|f(re^{i\theta}|^2)\, d\theta$$
$$\geqslant\frac{1}{4\pi}\int_0^{2\pi} log^+|f(re^{i\theta})|^2 d\theta=m(r,\infty).$$

Hence

$$C(f)\leqslant S(r,f)-T(r,f)\leqslant\frac{log2}{2}+C(f).$$

Theorem 1.1.8. $(\frac{1+\pi^2}{2})^2A(r,logf(z))\geqslant A(r,f(z)).$

Proof. First we note that to any $k>0$, the following inequality

$$\frac{1+\pi^2}{2}(k+\frac{1}{k})\geqslant 1+(log\ k)^2+\pi^2$$

holds. Hence

$$\frac{1+\pi^2}{2}(|f(z)|+\frac{1}{|f(z)|})\geqslant 1+(log|f(z)|)^2+\pi^2\geqslant 1+|logf(z)|^2,$$

consequently

$$(\frac{1+\pi^2}{2})^2\frac{|\frac{f'(z)}{f(z)}|}{(1+|logf(z)|^2)^2}\geqslant\frac{|f'(z)|^2}{(1+|f(z)|^2)^2}.$$

The assertion of the theorem follows from this and the definitions of $A(r,log\ f(z))$, and $A(r,f(z))$.

Theorem 1.1.9. *Let $f(z)$ be a meromorphic function in the finite complex plane with $f(0)\neq 0,\infty$ and $|f(0)|>\varepsilon$, $|\frac{1}{f(0)}|>\varepsilon(\varepsilon>0)$. Then there exists a constant $K(\varepsilon,f(0))$ depending on $f(0)$ and ε such that*

$$A(r,log\ f)\leqslant(\varepsilon+\frac{1}{\varepsilon})^2A(r,f)+\frac{\rho}{\rho-r}\{S(\rho,f)+K(\varepsilon,f(0))\}\quad (r<\rho<R).$$

$$\textit{Proof.}\ \pi A(r,\log f)=\iint_{|z|\leqslant r}\frac{\left|\frac{dw}{w}\right|^2}{(1+|\log w|^2)^2}\quad (w=f(z))$$
$$=\iint_{\substack{|w|\leqslant\varepsilon\\|z|\leqslant r}}\frac{|dw|^2}{(1+|\log w|^2)^2|w|^2}$$
$$+\iint_{\substack{\varepsilon<|w|<\frac{1}{\varepsilon}\\|z|\leqslant r}}\frac{|dw|^2}{(1+|\log w|^2)^2|w|^2}$$
$$+\iint_{\substack{|w|\geqslant\frac{1}{\varepsilon}\\|z|\leqslant r}}\frac{|dw|^2}{(1+|\log w|^2)^2|w|^2}. \tag{1.1.2}$$

The second term of the right side of (1. 1. 2) is

$$\iint_{\varepsilon<|w|<\frac{1}{\varepsilon}}\frac{|dw|^2}{(1+|\log w|^2)^2|w|^2}$$
$$=\iint_{\varepsilon<|w|<\frac{1}{\varepsilon}}\frac{|dw|^2}{(1+|w|^2)^2}\,\frac{(|w|+\frac{1}{|w|})^2}{(1+|\log w|^2)^2}$$
$$\leqslant\pi(\varepsilon+\frac{1}{\varepsilon})^2\,A(r,f).$$

And

$$\iint_{|w|\leqslant\varepsilon}+\iint_{|w|\geqslant\frac{1}{\varepsilon}}\frac{|dw|^2}{(1+|\log w|^2)|w|^2}<\pi\cdot n(r,w_{\varepsilon,r}).$$

where w_ε denotes, under fixed r, the point that maximizes the value of $n(r,w)$ when $|w|<\varepsilon$ or $|w|>1/\varepsilon$.

Thus

$$A(r,\log f)\geqslant(\varepsilon+\frac{1}{\varepsilon})^2A(r,f)+n(r,w_{\varepsilon,r}).$$

It follows that to any r and ρ satisfying $r<\rho<R$

$$n(r,w_{\varepsilon,r})\leqslant\frac{\rho}{\rho-r}\int_r^\rho\frac{n(r,w_{\varepsilon,r})}{t}dt\leqslant\frac{\rho}{\rho-r}N(\rho,w_{\varepsilon,r}).$$

Also by Theorem 1. 1. 4

$$N(\rho,w_{\varepsilon,r})\leqslant S(\rho,f)+\frac{1}{2}\log\left(1+\left|\frac{1+\overline{w}_{\varepsilon,r}f(0)}{f(0)-w_{\varepsilon,r}}\right|^2\right).$$

Furthermore, when $|w_{\varepsilon,r}|\leqslant\varepsilon$

$$\left|\frac{1+\overline{w}_{\varepsilon,r}f(0)}{f(0)-w_{\varepsilon,r}}\right|\leqslant\frac{1+\varepsilon|f(0)|}{|f(0)|-\varepsilon};$$

and, when $|w_{\varepsilon,r}|\geqslant\frac{1}{\varepsilon}$

$$\left|\frac{1+\overline{w}_{\varepsilon,r}f(0)}{f(0)-w_{\varepsilon,r}}\right|\leqslant\frac{\varepsilon+|f(0)|}{1-\varepsilon|f(0)|}.$$

Now we choose

$$K=\max\left\{\frac{1}{2}\log\left[1+\left(\frac{1+\varepsilon|f(0)|}{|f(0)|-\varepsilon}\right)^2\right],\ \frac{1}{2}\log\left[1+\left(\frac{\varepsilon+|f(0)|}{1-\varepsilon|f(0)|}\right)^2\right]\right\}.$$

the assertion of the theorem follows.

1.5. THE PROPERTIES OF $m\left(r,\frac{f'}{f}\right)$

Let $f(z)$ be a function meromorphic in $|z|<R$ with $f(0)\neq 0,\infty$; namely

$$f(z)=c_0+c_kz^k+\cdots+c_nz^n+\cdots,$$
$$c_0=f(0)\neq 0,\infty.$$

It follows from the definition of $A(r,f)$, that:

$$A(r,\log f(z))=\frac{1}{\pi}\int_0^r\int_0^{2\pi}\frac{\left|\frac{f'(z)}{f(z)}\right|^2}{(1+|\log f(z)|^2)^2}\,t\,dt\,d\theta\ (z=te^{i\theta}).$$

Let

$$\log f(z)=w_1(z)=u_1(z)+iv_1(z)$$

with $-\pi<v_1\leqslant\pi$. Then to any ρ, r satisfying $R>\rho>r$

$$\frac{A(\rho,w_1)}{2(\rho-r)}+\frac{\rho+r}{2}$$

$$\geqslant\frac{1}{2\pi(\rho-r)}\int_r^\rho\int_0^{2\pi}\frac{\left|\frac{f'(te^{i\theta})}{f(te^{i\theta})}\right|^2+1}{(1+|w_1(te^{i\theta})|^2)^2}\,t\,dt\,d\theta.$$

Thus by the corollary of Theorem 1.1.2 and the above, we have

$$\log\left\{\frac{A(\rho,w_1)}{2(\rho-r)}+\frac{\rho+r}{2}\right\}$$

$$\geqslant\log\left\{\frac{r}{2\pi(\rho-r)}\int_r^\rho\int_0^{2\pi}\frac{1+\left|\frac{f'}{f}\right|^2}{(1+|w_1|^2)}\,dt\,d\theta\right\}$$

$$\geqslant \log r + \frac{1}{2\pi(\rho - r)} \int_r^\rho \int_0^{2\pi} \log \frac{1+\left|\frac{f'}{f}\right|^2}{(1+|w_1|^2)^2}\, dt\, d\theta. \tag{1.1.3}$$

Also

$$\frac{1}{2\pi(\rho-r)} \int_r^\rho \int_0^{2\pi} \log \frac{1+\left|\frac{f'}{f}\right|}{(1+|w_1|^2)^2}\, dt\, d\theta$$
$$= \frac{1}{\rho-r}\int_r^\rho \frac{1}{2\pi}\int_0^{2\pi} \log\left(1+\left|\frac{f'}{f}\right|^2\right) dt\, d\theta$$
$$-\frac{1}{\rho-r}\int_r^\rho \frac{1}{2\pi}\int_0^{2\pi} \log(1+|w_1|^2)^2 dt\, d\theta. \tag{1.1.4}$$

Since

$$|w_1|^2 = |u_1|^2 + |v_1|^2 \leqslant |u_1|^2 + \pi^2,$$

it follows that

$$\frac{1}{\rho-r}\int_r^\rho \frac{1}{2\pi}\int_0^{2\pi} \log(1+|w_1|^2)^2 dt\, d\theta$$
$$\leqslant \frac{2}{\rho-r}\int_r^\rho \frac{1}{2\pi}\int_0^{2\pi} \log(1+\pi^2+|u_1|^2) dt\, d\theta$$
$$\leqslant \frac{2}{\rho-r}\int_r^\rho \frac{1}{2\pi}\int_0^{2\pi} \{\log 2 + \log(1+\pi^2) + \log^+ |u_1|^2\} dt\, d\theta$$
$$= 2\log 2(1+\pi^2) + \frac{4}{\rho-r}\int_r^\rho \frac{1}{2\pi}\int_0^{2\pi} \log^+ |u_1|\, dt\, d\theta. \tag{1.1.5}$$

Now $u_1 = \log|f|$, hence

$$\frac{1}{2\pi}\int_0^{2\pi} \log^+ |u_1|\, d\theta$$
$$\leqslant \frac{1}{2\pi}\int_0^{2\pi} \log(|u_1|+1) d\theta$$
$$\leqslant \log \frac{1}{2\pi}\int_0^{2\pi} (|u_1|+1) d\theta$$
$$\leqslant \log \frac{1}{2\pi}\int_0^{2\pi} (\log^+ |f| + \log^+ \frac{1}{|f|} + 1) d\theta$$
$$\leqslant \log^+ \left\{ \frac{1}{2\pi}\int_0^{2\pi} \frac{1}{2}\left[\log(|f|^2+1) + \log\left(\frac{1}{|f|^2}+1\right)\right] d\theta \right\} + \log 2$$
$$= \log\left\{ q(t,f) + q\left(t, \frac{1}{f}\right)\right\} + \log 2$$
$$\leqslant \log^+ \left\{ 2S(t,f) + \frac{1}{2}(1+|f(0)|^2) + \frac{1}{2}\log\left(1+\frac{1}{|f(0)|^2}\right)\right\} + 2\log 2$$

$$\leqslant log^+ S(t,f) + log^+ log \frac{1+|f(0)|^2}{|f(0)|} + 3\ log 2. \text{①}$$

We note that $S(r,f)$ is a monotonic increasing function of r, hence

$$\frac{4}{\rho - r}\int_r^\rho \frac{1}{2\pi}\int_0^{2\pi} log^+ |u_1|\ d\theta\ dt$$

$$\leqslant 4\left\{ log^+ S(\rho,f) + log^+ log \frac{1+|f(0)|^2}{|f(0)|} + 3\ log 2\right\}. \tag{1.1.6}$$

It follows from (1.1.3), (1.1.4), (1.1.5), and (1.1.6) that

$$log\left\{\frac{A(\rho,w_1)}{2(\rho-r)} + \frac{\rho+r}{2}\right\} + log^+\frac{1}{r}$$

$$+ 2log 2(1+\pi^2) + 4log^+ S(\rho,f) + 12 log 2$$

$$+ 4log^+ log\frac{1+|f(0)|^2}{|f(0)|} \geqslant \frac{2}{\rho - r}\int_r^\rho q\left(t,\frac{f'}{f}\right)dt. \tag{1.1.7}$$

Thus by Theorems 1.1.5 and 1.1.6 the monotonic increasing of $S(r,f)$ and $N(r,f)$, we obtain for any $\rho > t > r$

$$q\left(t,\frac{f'}{f}\right) + N\left(\rho,\frac{f'}{f}\right) \geqslant q\left(r,\frac{f'}{f}\right) + N\left(r,\frac{f'}{f}\right),$$

hence

$$\frac{1}{\rho - r}\int_r^\rho q\left(t,\frac{f'}{f}\right)dt \geqslant q\left(r,\frac{f'}{f}\right) - \left\{N\left(\rho,\frac{f'}{f}\right) - N\left(r,\frac{f'}{f}\right)\right\}. \tag{1.1.8}$$

By Theorem 1.1.9 (with the assumption that $\varepsilon < |f(0)|$, $\varepsilon < \frac{1}{|f(0)|}$, $\varepsilon > 0$) we have, for any ρ' with $R > \rho' > \rho$

$$A(\rho,w_1) \leqslant \left(\varepsilon + \frac{1}{\varepsilon}\right)^2 A(\rho,f) + \frac{\rho'}{\rho' - \rho}\{S(\rho',f) + K(\varepsilon,f(0))\}.$$

Substituting

$$A(\rho,f) \leqslant \frac{\rho'}{\rho' - \rho}\int_\rho^{\rho'} \frac{A(t,f)}{t}dt$$

into the above inequality, we obtain

$$A(\rho,w_1) \leqslant \frac{\rho'}{\rho' - \rho}\{[(\varepsilon + \frac{1}{\varepsilon})^2 + 1]S(\rho',f) + K(\varepsilon,f(0))\}. \tag{1.1.9}$$

Hence

---

① Here Theorem 1.1.4 and the fact $S(r.f) = S(r, L_a(f))$ are used, where $L_a(z)$ denotes transformation of rotation on the Riemann sphere.

$$\log^+\frac{1}{2(\rho-r)}+\log^+\frac{\rho+r}{2}+2\log 2+\log^+\frac{\rho'}{\rho'-\rho}+\log\left[\left(\varepsilon+\frac{1}{\varepsilon}\right)^2+1\right]$$

$$+\log^+ S(\rho',f)+\log^+ K(\varepsilon,f(0))\geqslant \log\left\{\frac{A(\rho,w_1)}{2(\rho-r)}+\frac{\rho+r}{2}\right\}. \quad (1.1.10)$$

Combining (1.1.7), (1.1.8), and (1.1.10), we get

$$\log^+\frac{1}{r}+\log 2^{16}(1+\pi^2)^2+2\log^+\rho'+\log^+\frac{1}{\rho'-\rho}+$$

$$\log^+\frac{1}{\rho-r}+5\log^+ S(\rho',f)+4\log^+\log\frac{1+|f(0)|^2}{|f(0)}+$$

$$\log\left[\left(\varepsilon+\frac{1}{\varepsilon}\right)^2+1\right]+\log^+ K(\varepsilon,f(0))$$

$$+2\left\{N\left(\rho,\frac{f'}{f}\right)-N\left(r,\frac{f'}{f}\right)\right\}>2q\left(r,\frac{f'}{f}\right). \quad (1.1.11)$$

Noting the poles of $\frac{f'(z)}{f(z)}$ come from the poles and zeros of f and all are simple poles. We have

$$N\left(r,\frac{f'}{f}\right)\leqslant N(r,f)+N\left(r,\frac{1}{f}\right)\equiv N(r).$$

Let $r<\rho$, then

$$N\left(\rho,\frac{f'}{f}\right)-N\left(r,\frac{f'}{f}\right)\leqslant N(\rho)-N(r). \quad (1.1.12)$$

Now assume r and ρ are given such that

$$\rho-r=\frac{\rho'-r}{S(\rho',f)+2}\,\frac{r}{\rho'}. \quad (1.1.13)$$

Then, for $r<\rho<\rho'$

$$\rho-r<\frac{\rho'-r}{S(\rho',f)+2}.$$

According to Theorem 1.1.6 $N(r)$ is a convex function of $\log r$, hence

$$\frac{N(\rho)-N(r)}{\log\rho-\log r}\leqslant\frac{N(\rho')-N(r)}{\log\rho'-\log r}\leqslant\frac{N(\rho')}{\log\rho'-\log r}.$$

Since

$$\log\rho-\log r=\log\left(1+\frac{\rho-r}{r}\right)\leqslant\frac{\rho-r}{r},$$

$$\log\rho'-\log r=-\log\left(1-\frac{\rho'-r}{\rho'}\right)\geqslant\frac{\rho'-r}{\rho'},$$

we can derive from the above that

$$N(\rho)-N(r)\leqslant\frac{\rho'}{r}\frac{\rho-r}{\rho'-r}N(\rho').$$

Thus by Theorem 1.1.4 and (1.1.13) we obtain

$$N(\rho)-N(r)$$

$$\leqslant\frac{2S(\rho',f)+\frac{1}{2}\log(1+|f(0)|^2)+\frac{1}{2}\log\left(1+\frac{1}{|f(0)|^2}\right)}{S(\rho',f)+2}$$

$$\leqslant 2+\frac{1}{2}\log\frac{1+|f(0)|^2}{|f(0)|}. \tag{1.1.14}$$

Also by (1.1.13) we have

$$\rho'-\rho=(\rho'-r)\left(1-\frac{r}{\rho'[S(\rho',f)+2]}\right)\geqslant\frac{\rho'-r}{2},$$

and hence

$$\log^+\frac{1}{\rho'-\rho}\leqslant\log 2+\log^+\frac{1}{\rho'-r}. \tag{1.1.15}$$

Again by (1.1.13)

$$\log^+\frac{1}{\rho-r}\leqslant\log^+\frac{S(\rho',f)+2}{\rho'-r}+\log\frac{\rho'}{r}$$

$$\leqslant 2\log 2+\log^+\frac{1}{\rho'-r}+\log^+\rho'+\log^+\frac{1}{r}+\log^+S(\rho',f). \tag{1.1.16}$$

substituting (1.1.12), (1.1.14), (1.1.15) and (1.1.16) into (1.1.11) and eliminating the term that contains ρ, and then replacing ρ' by ρ, we get

$$q\left(r,\frac{f'}{f}\right)<3\log^+S(\rho,f)+\frac{3}{2}\log^+\rho+\log^+\frac{1}{r}$$

$$+\log^+\frac{1}{\rho-r}+\log e^2 2^{\frac{19}{2}}(1+\pi^2)+\frac{1}{2}\log\left[\left(\varepsilon+\frac{1}{\varepsilon}\right)^2+1\right]$$

$$+\frac{1}{2}\log^+K(\varepsilon,f(0))+\frac{1}{2}\log\frac{1+|f(0)|^2}{|f(0)|}$$

$$+2\log^+\log\frac{1+|f(0)|^2}{|f(0)|}. \tag{1.1.17}$$

When $|f(0)|\geqslant 1$, choose $\varepsilon=\frac{1}{2|f(0)|}$, and when $|f(0)|\leqslant 1$, choose $\varepsilon=\frac{1}{2}|f(0)|$. Then when $|f(0)|\geqslant 1$,

$$\left(\varepsilon+\frac{1}{\varepsilon}\right)^2\leqslant\frac{25}{4}|f(0)|^2,$$

$$K(\varepsilon,f(0))\leqslant\frac{\log 10}{2}+\log|f(0)|.$$

And when $|f(0)|\leqslant 1$,

$$\left(\varepsilon+\frac{1}{\varepsilon}\right)^2\leqslant\frac{25}{4}\frac{1}{|f(0)|^2},$$

$$K(\varepsilon, f(0)) \leqslant \frac{\log 10}{2} + \log \frac{1}{|f(0)|}.$$

Thus from (1. 1. 17) we arrive at the following theorem.

Theorem 1. 1. 10. (Nevanlinna). *Let $f(z)$ be a function meromorphic in $|z|<R$ with $f(0)=c_0 \neq 0, \infty$. Then for $R>\rho>r$*

$$m\left(r, \frac{f'}{f}\right) > q\left(r, \frac{f'}{f}\right) < 3\log^+ S(\rho, f)$$
$$+\log^+ \frac{1}{\rho - r} + \frac{3}{2}\log^+ \rho + \log^+ \frac{1}{r} + p(|c_0|) + k,$$

where

$$p(|c_0|) = \frac{3}{2}\left(\log^+ |c_0| + \log^+ \frac{1}{|c_0|}\right)$$
$$+\frac{5}{2}\left(\log^+ \log |c_0| + \log^+ \log^+ \frac{1}{|c_0|}\right) \tag{1. 1. 18}$$

and

$$k = 2^{\frac{19}{2}} \cdot 3^{\frac{5}{2}} \cdot 5e^2 (1+\pi^2) \sqrt{\log \sqrt{10}}. \tag{1. 1. 19}$$

For $f(0)=0$ or ∞, i. e. $f(z)=z^d g(z)$, $d \geqslant 0$, we have

$$m\left(r, \frac{f'}{f}\right) \leqslant m\left(r, \frac{d}{z}\right) + m\left(r, \frac{g'}{g}\right) + \log 2$$
$$\leqslant \log^+ |d| + \log^+ \frac{1}{r} + m\left(r, \frac{g'}{g}\right) + \log 2. \tag{1. 1. 20}$$

From $g = \frac{f}{z^d}$, we have

$$q(r, g) \leqslant q\left(r, \frac{1}{z^d}\right) + q(r, f),$$

$$N(r, g) \leqslant N\left(r, \frac{1}{z^d}\right) N(r, f).$$

Hence by Theorem 1. 1. 4 for $d>0$ and, hence, for $d<0$ the following result holds:

$$S(r, g) < S\left(r, \frac{1}{z}\right) + S(r, f). \tag{1. 1. 21}$$

But

$$S\left(r, \frac{1}{z^d}\right) = \frac{1}{2}(1 + r^{|2d|}). \tag{1. 1. 22}$$

Hence it follows from (1. 1. 21), (1. 1. 22) and Theorem 1. 1. 10 that

$$m\left(r,\frac{g'}{g}\right)<q\left(r,\frac{g'}{g}\right)<3\log^+\left\{\frac{1}{2}\log(1+\rho^{|2d|})+S(\rho,f)\right\}$$

$$+\log^+\frac{1}{\rho-r}+\frac{3}{2}\log^+\rho+\log^+\frac{1}{r}+p(|(g(0)|)+k. \qquad (1.1.23)$$

When $d<0$, we have

$$\frac{1}{2}\log(1+r^{2d})<|d|\left\{\log^+r+\log^+\frac{1}{r}\right\}+\frac{\log 2}{2}.$$

Thus, by combining (1.1.20), (1.1.23), and the fact that $g(0)(=c_{-d}$ is the first nonzero coefficient in the Laurent expansion of $f(z)$ around $z=0$, we obtain the following conclusion.

Theorem 1.1.11. *Let $f(z)$ be meromorphic in $|z|<R$ with its Laurent expansion around $z=0$ being*

$$\frac{c_{-d}}{z^d}+\frac{c_{-(d-1)}}{z^{d-1}}+\cdots+\frac{c_{-1}}{z}+c_0+c_1z+\cdots+c_nz^n+\cdots.$$

Then, when $d\gtreqless 0$ with $c_{-d}\neq 0$

$$m\left(r,\frac{f'}{f}\right)<3\log^+S(\rho,f)+\log^+\frac{1}{\rho-r}$$

$$+\frac{3}{2}\log^+\rho+2\log^+\frac{1}{r}$$

$$+3\left\{\log^+\log^+r+\log^+\log^+\frac{1}{r}\right\}$$

$$+4\log^+|d|+p(|C_{-d}|)+k_1,$$

where

$$p(|C_{-d}|)=\frac{3}{2}\left\{\log^+|C_{-d}|+\log^+\frac{1}{|C_{-d}|}\right\}$$

$$+\frac{5}{2}\left\{\log^+\log^+|C_{-d}|+\log^+\log^+\frac{1}{|C_{-d}|}\right\}, \qquad (1.1.24)$$

$$k_1=k+\log 2^{10}. \qquad (1.1.25)$$

1.6. THE RELATIONSHIPS AMONG $T(r,f)$, $T(r,f')$, AND $N(r,f)$ ETC.

Let $f(z)$ be meromorphic in $|z|<R$ with

$$\frac{c_{-d}}{z^d}+\frac{c_{-(d-1)}}{z^{d-1}}+\cdots+\frac{c_{-1}}{z}+c_0+c_1z$$

$$+\cdots+c_nz^n+\cdots(d\gtreqless 0),$$

as its Laurent expansion around $z=0$.

Let $w_1, w_2, \cdots w_q$ be q distinct finite values.

From $f' = f \cdot \frac{f'}{f}$, it follows that

$$T(r,f') = m(r,f') + N(r,f')$$
$$\leqslant m(r,f) + m\left(r, \frac{f'}{f}\right) + N(r,f'). \tag{1.1.26}$$

We shall use the property

$$T(r,f) = T\left(r, \frac{1}{f}\right) + \log|c_{-d}|,$$

to get an estimation of $T(r,f')$.

First of all, we have

$$T(r,f') = T\left(r, \frac{1}{f'}\right) + \log|\beta_{-d}|$$
$$= N\left(r, \frac{1}{f'}\right) + m\left(r, \frac{1}{f'}\right) + \log|\beta_{-d}|, \tag{1.1.27}$$

where

$$\beta_{-d} = \begin{cases} -dc_{-d}, & \text{when } d \neq 0, \\ kc_k, & \text{when } d = 0 \end{cases}$$

and c_k is the first nonzero coefficient in the Laurent expansion expressed above.

Put

$$F(z) = \frac{F(z)f'(z)}{f'(z)}, \quad F(z) = \sum_{v=1}^{q} \frac{1}{f(z) - w_v},$$

then

$$m(r,F) \leqslant m\left(r, \frac{1}{f'}\right) + m\left(r, \sum_{v=1}^{q} \frac{f'}{f-w}\right) \tag{1.1.28}$$

or

$$F(z) = \frac{1}{f(z) - w_\mu}\left[1 + \sum_{v \neq \mu} \frac{f(z) - w_\mu}{f(z) - w_v}\right], \quad \mu \in \{1, 2, 3, \cdots, q\}. \tag{1.1.29}$$

Set

$$\delta(w) = \min\{|w_h - w_k| : h \neq k, \ 1 \leqslant h, k \leqslant q\}.$$

Then, for any z satisfying inequality

$$|f(z) - w_\mu| < \frac{\delta(w)}{2q} \leqslant \frac{\delta(w)}{4} \tag{1.1.30}$$

we have

$$|f(z) - w_v| \geqslant |w_\mu - w_v| - |f(z) - w_\mu| > \delta(w) - \frac{\delta(w)}{2q} \geqslant \frac{3}{4}\delta(w) \ (v \neq \mu).$$

Hence

$$\sum_{v\neq\mu}\left|\frac{f(z)-w_\mu}{f(z)-w_v}\right|<q\cdot\frac{2}{3q}=\frac{2}{3}.$$

Consequently

$$\left|1+\sum_{v\neq\mu}\frac{f(z)-w_\mu}{f(z)-w_v}\right|>\frac{1}{3}.$$

Thus for any z satisfying (1. 1. 30), from (1. 1. 29) we have

$$\log^+|F(z)|>\log^+\left|\frac{1}{f(z)-w_\mu}\right|-\log 3.$$

Now the arcs on $|z|=r$ that satisfy (1. 1. 30) for $\mu=1,2,\cdots,q$ are disjoint from each other.

Thus

$$m(r,F)\geqslant\frac{1}{2\pi}\sum_{\mu=1}^{q}\int_{|f-w_\mu|<\frac{\delta(w)}{2q}}\log^+|F(re^{i\theta})|d\theta$$

$$>\frac{1}{2\pi}\sum_{\mu=1}^{q}\int_{|f-w_\mu|<\frac{\delta(w)}{2q}}\log^+\left|\frac{1}{f(re^{i\theta})-w_\mu}\right|d\theta-\log 3. \qquad (1.\ 1.\ 31)$$

While

$$\frac{1}{2\pi}\int_{|f-w_\mu|<\frac{\delta(w)}{2q}}\log^+\left|\frac{1}{f(re^{i\theta})-w_\mu}\right|d\theta$$

$$=\frac{1}{2\pi}\int_0^{2\pi}\log^+\left|\frac{1}{f(re^{i\theta})-w_\mu}\right|d\theta$$

$$-\frac{1}{2\pi}\int_{|f-w_\mu|\geqslant\frac{\delta(w)}{2q}}\log^+\left|\frac{1}{f(re^{i\theta}-w_\mu}\right|d\theta$$

$$\geqslant m\left(r,\frac{1}{f-w_\mu}\right)-\log^+\frac{2q}{\delta(w)}. \qquad (1.\ 1.\ 32)$$

Therefore, it follows from (1. 1. 31) and (1. 1. 32) that

$$m(r,F)>\sum_{v=1}^{q}m\left(r,\frac{1}{f-w_v}\right)-q\log^+\frac{2q}{\delta(w)}-\log 3. \qquad (1.\ 1.\ 33)$$

Combining this with

$$T(r,f)=T(r,\frac{1}{f})+\log|c_{-d}|,$$

and

$$T(r,f)\leqslant T(r,f-w_v)+\log^+|w_v|+\log 2,$$

we obtain

$$\sum_{v=1}^{q} m\left(r,\frac{1}{f-w_v}\right) \geqslant qT(r,f) - \sum_{v=1}^{q} N\left(r,\frac{1}{f-w_v}\right)$$
$$-q\log 2 - \sum_{v=1}^{q}\log^+|w_v| - \log|G_0|, \qquad (1.1.34)$$

where G_0 is the first nonzero coefficient of the Laurent expansion of $G(z)=\prod_{v=1}^{q}(f(z)-w_v)$ around $z=0$.

From (1.1.27), (1.1.28), (1.1.33), and (1.1.34), we get

$$T(r,f') \geqslant N\left(r,\frac{1}{f'}\right) + q\,T(r,f)$$
$$-\sum_{v=1}^{q} N\left(r,\frac{1}{f-w_v}\right) - m\left(r,\sum_{v=1}^{q}\frac{f'}{f-w_v}\right)$$
$$+\log|a_{-k}| - q\log^+\frac{2q}{\delta(w)} - \log 3 - q\log 2$$
$$-\sum_{v=1}^{q}\log^+|w_v| - \log|G_0|. \qquad (1.1.35)$$

Furthermore since

$$m\left(r\sum_{\mu=1}^{q}\frac{f'}{f-w_\mu}\right) = m\left(r,\frac{G'}{G}\right),$$

and by Theorem 1.1.7, we obtain

$$m\left(r,\frac{G'}{G}\right) < 3\log^+ S(\rho,G) + \log^+\frac{1}{\rho-r} + \frac{3}{2}\log^+\rho$$
$$+2\log^+\frac{1}{r} + 3\left\{\log^+\log^+ r + \log^+\log^+\frac{1}{r}\right\}$$
$$+4\log^+|m_G| + p(|G_0|) + k_1, \qquad (1.1.36)$$

where

$$p(|G_0|) = \frac{3}{2}\left\{\log^+|G_0| + \log^+\left|\frac{1}{G_0}\right|\right\}$$
$$+\frac{5}{2}\left\{\log^+\log^+|G_0| + \log^+\log^+\left|\frac{1}{G_0}\right|\right\},$$
$$k_1 = k + \log 2^{10},$$

and m_G denote the multiplicity of $G(0)=0$ or ∞; $m_G=0$ if $G(0)\neq 0,\infty$.

It follows from Theorem 1.1.7 (regardless of $G(0)\neq\infty$ or $G(0)=\infty$) that

$$\log^+ S(\rho,G) < \log^+ T(\rho,G) + \log^+\log^+\left|\frac{1}{G_0}\right| + \log 3. \qquad (1.1.37)$$

Now as it is easily shown that

$$T(r,fg)\leqslant T(r,f)+T(r,g),$$
$$T(r,f+g)\leqslant T(r,f)+T(r,g)+\log 2.$$

We have

$$T(\rho,G)\leqslant\sum_{v=1}^{q}T(\rho,f-w_v),$$
$$T(\rho,f-w_v)\leqslant T(\rho,f)+\log^+|w_v|+\log 2,$$
$$T(\rho,G)\leqslant qT(\rho,f)+q\log 2+\sum_{v=1}^{q}\log^+|w_v|.$$

Hence

$$\log^+S(\rho,G)\leqslant 2\log q+\log^+T(\rho,f)$$
$$+\log^+\left(\sum_{v=1}^{q}\log^+|w_v|\right)+2\log 3+\log^+\log^+\left|\frac{1}{C_0}\right|. \qquad (1.1.38)$$

Thus by combining (1.1.26), (1.1.35), (1.1.36), (1.1.37), (1.1.38), and applying Theorem 1.1.11 to $m(r,\frac{f'}{f})$, we can obtain the following.

Theorem 1.1.12.

$$qT(r,f)-\sum_{v=1}^{q}N\left(r,\frac{1}{f-w}\right)$$
$$+N\left(r,\frac{1}{f'}\right)-\{3\log^+T(\rho,f)$$
$$+\log^+\frac{1}{\rho-r}+\frac{3}{2}\log\rho+2\log^+\frac{1}{r}+3\left(\log^+\log^+r+\log^+\log^+\frac{1}{r}\right)$$
$$+\sum_{v=1}^{q}\log^+|w_v|+3\log^+\left(\sum_{v=1}^{q}\log^+|w_v|\right)$$
$$+3\log^+\log^+\left|\frac{1}{G_0}\right|+\log|G_0|+p(|G_0|)+6\log q$$
$$+q\log^+\frac{2q}{\delta(\omega)}+4\log^+|m_G|+7\log 3+q\log 2+k_1\}$$
$$+\log|a_{-l}|\leqslant T(r,f')\leqslant N(r,f')+m(r,f)$$
$$+3\log^+T(\rho,f)+3\log^+\log^+\left|\frac{1}{c_{-l}}\right|+3\log 3$$
$$+\log^+\frac{1}{\rho-r}+\frac{3}{2}\log^+\rho+2\log^+\frac{1}{r}+3\{\log^+\log^+r$$

$$+ log^+ log^+ \frac{1}{r} \} + 4 log^+ |d| + p(|c_{-d}|) + k_1.$$

From

$$N(r, f') + m(r, f) \leqslant N(r, f') - 2N(r, f) + 2N(r, f) + 2m(r, f),$$

and from the above theorem it follows that

$$(q-2)T(r, f) \leqslant \sum_{v=1}^{q} N\left(r, \frac{1}{f - w_v}\right) - N_1(r) + D(\rho, r),$$

$$N_1(r) = 2N(r, f) - N(r, f') + N\left(r, \frac{1}{f'}\right), \qquad (1.1.39)$$

where

$$D(\rho, r) = 6 log^+ T(\rho, f) + 2 log^+ \frac{1}{\rho - r} + 3 log^+ \rho$$

$$+ 4 log^+ \frac{1}{r} + 6\left(log^+ log^+ r + log^+ log^+ \frac{1}{r} \right)$$

$$+ 3 log^+ log^+ \left| \frac{1}{c_{-d}} \right| + 4 log^+ |m| + p(|c_{-d}|)$$

$$+ \sum_{v=1}^{q} log^+ |w_v| + 3 log^+ \left(\sum_{v=1}^{q} log^+ |w_v| \right)$$

$$+ 3 log^+ log^+ \left| \frac{1}{G_0} \right| + log |G_0| + p(|G_0|) + 6 log\ q$$

$$+ q log^+ \frac{2q}{\delta(w)} + 4 log^+ |m_G| + 10 log 3$$

$$+ q log 2 + 2k_1 - log |a_{-d}|. \qquad (1.1.40)$$

Moreover, if, in Theorem 1.1.12, by adding $T(r, f) - N(r, f)$ and $m(r, f) = m(r, f) + 2N(r, f) - 2N(r, f)$ to the left side and right side of $T(r, f')$, respectively, then by a simple calculation then

$$(q-1)T(r, f) \leqslant \sum_{v=1}^{q} N\left(r, \frac{1}{f - w_v}\right) + N(r, f) - N_1(r) + D(\rho, r).$$

Thus the following important result is obtained.

Theorem 1.1.13.(Littlewood-Collingwood-Nevanlinna theorem or Nevanlinna's second fundamental theorem). Let $f(z)$ be meromorphic in $|z| < R$ and $w_1, w_2 \cdots, w_q$ be q distinct values(possibly may include ∞). Then, to $R > \rho > r > 0$, the following inequality holds:

$$(q-2)T(r, f) < \sum_{v=1}^{q} N\left(r, \frac{1}{f - w_v}\right) - N_1(r) + D(\rho, r).$$

Here, according to (1. 1. 40), *when* $w_1, \cdots, w_q$ *all are finite, the term* $D(\rho, r)$ *stands as it is; when one of the w' s, say* $w_q = \infty$, *then in* (1. 1. 40) $q-1$ *will replace q in the upper index of the summations that concern the w' s.*

Corollary. *Let* $f(z)$ *be meromorphic in* $|z|<R$ *and* $w_1, w_2, \cdots, w_q$ *be* $q (q \geqslant 2)$ *distinct values* (∞ *is allowed*). *If* $f(z) \neq w_1, w_2, \cdots, w_q$ *in* $|z|<R$, *then for* $R>\rho>r>0$

$$(q-2)T(r,f)<D(\rho,r),$$

where $D(\rho,r)$ *is defined as in Theorem* 1. 1. 12.

1. 7. THE INFINITE PRODUCT REPERESENTATION OF A MEROMORPHIC FUNCTION

We shall consider the representation of a finite order meromorphic functions, in the plane, by a infinite product formed with their zeros and poles. The following result is standard.

Theorem 1. 1. 14. *Let* $f(z)$ *be a meromorphic function of order* λ. *Then*

$$f(z)=z^k e^{P(z)} \lim_{\rho\to\infty} \prod_{\substack{|a_\mu|\leqslant\rho \\ |b_\nu|\leqslant\rho}} \left\{ \frac{\left(1-\frac{z}{a_\mu}\right) exp\left(\frac{z}{a_\mu}+\cdots+\frac{z^p}{pa_\mu^p}\right)}{\left(1-\frac{z}{b_\nu}\right) exp\left(\frac{z}{b_\nu}+\cdots+\frac{z^p}{pb_\mu^p}\right)} \right\}, \tag{1. 1. 41}$$

where p is the the smallest integer (including zero) that satisfies $p+1>\lambda$, *k is an integer (negative or positive) or zero,* $P(z)$ *is a polynomial of degree* $\leqslant p$, $\prod$ *represents a product, within the disk:* $|z|<\rho$, *formed with all the zeros* a_μ *and poles* b_ν; *zeros or poles with multiplicity m will be repeated m times in the product.*

The right side of expression (1. 1. 41) *converges uniformaly in general sense in any bounded domain in D not containing zeros or poles of* $f(z)$. ①

Particularly, if for the integer p, $\lim_{r\to\infty} \int_{r_0}^{r} \frac{T(t,f)}{t^{p+2}} dt$ *is finite, then*

---

① Converges uniformly in general sense in *D* means it converges uniformly in any closed subregion of *D*.

$$f(z)=z^k e^{P(z)}\frac{\lim\limits_{\rho\to\infty}\prod\limits_{|a_\mu|\leqslant\rho}\left(1-\frac{z}{a_\mu}\right)exp(\frac{z}{a_\mu}+\cdots+\frac{z^p}{pa_\mu^p})}{\lim\limits_{\rho\to\infty}\prod\limits_{|b_\nu|\leqslant\rho}\left(1-\frac{z}{b_\nu}\right)exp(\frac{z}{b_\nu}+\cdots+\frac{z^p}{pb_\nu^p})}. \tag{1. 1. 42}$$

Proof. By assumption

$$\overline{\lim_{r\to\infty}}\frac{\log T(r,f)}{\log r}=\lambda.$$

It means that to any given $\varepsilon>0$, there corresponds a $r_0(\varepsilon)$ such that whenever $r>r_0(\varepsilon)$,

$$T(r,f)<r^{\lambda+\varepsilon}.$$

Now $p+1>\lambda$, so ε can be chosen satisfying $p+1>\lambda+\varepsilon$. Thus

$$\lim_{r\to\infty}\frac{T(r,f)}{r^{p+1}}=0.$$

Consider the case that $f(0)\neq 0$ or ∞. Then for $f(z)\neq 0,\infty$, by Poisson-Jensen formula, we have

$$\log f(z)=\frac{1}{2\pi}\int_0^{2\pi}\log|f(\zeta)|\frac{\zeta+z}{\zeta-z}\,d\theta-\sum_{|a_\mu|\leqslant\rho}\log\frac{\bar a_\mu z-\rho^2}{\rho(z-a_\mu)}$$
$$+\sum_{|b_\nu|\leqslant\rho}\log\frac{\bar b_\nu z-\rho^2}{\rho(z-b_\nu)}+ic\ (\zeta=\rho e^{i\theta}). \tag{1. 1. 43}$$

Taking derivative with respect to z, we obtain from above

$$\frac{d\log f(z)}{dz}=\frac{1}{2\pi}\int_0^{2\pi}\log|f(\zeta)|\frac{2\zeta}{(\zeta-z)^2}d\theta$$
$$-\sum_{|a_\mu|\leqslant\rho}\frac{\bar a_\mu}{\bar a_\mu z-\rho^2}+\sum_{|b_\nu|\leqslant\rho}\frac{\bar b_\nu}{\bar b_\nu z-\rho^2}$$
$$+\sum_{|a_\mu|\leqslant\rho}\frac{1}{z-a_\mu}-\sum_{|b_\nu|\leqslant\rho}\frac{1}{z-b_\nu}. \tag{1. 1. 44}$$

By taking the p-th derivative of the above, we have

$$\frac{d^{p+1}}{dz^{p+1}}\log(z)=\frac{(p+1)!}{2\pi}\int_0^{2\pi}\log|f(\zeta)|\frac{2\zeta}{(\zeta-z)^{p+2}}d\theta$$
$$-(-1)^p\cdot p!\left[\sum_{|a_\mu|\leqslant\rho}\left(\frac{\bar a_\mu}{\bar a_\mu z-\rho_2}\right)^{p+1}-\sum_{|b_\nu|\leqslant\rho}\left(\frac{\bar b_\mu}{\bar b\mu z-\rho^2}\right)^{p+1}\right]$$
$$+(-1)^p\cdot p!\left[\sum_{|a_\mu|\leqslant\rho}\frac{1}{(z-a_\mu)^{p+1}}-\sum_{|b_\nu|\leqslant\rho}\frac{1}{(z-b_\nu)^{p+1}}\right]. \tag{1. 1. 45}$$

In order to derive the limit of the right side of (1. 1. 45) when $\rho\to\infty$, we denote the right side of the above equation as $\equiv I_1+I_2+I_3$.

First of all, for $|z|<r$,

$$|I_1|\leqslant\frac{(p+1)!}{2\pi}\int_0^{2\pi}|\log|f(\zeta)||\frac{2\rho}{(\rho-r)^{p+2}}d\theta,$$

but

$$\frac{1}{2\pi}\int_0^{2\pi}|\log|f(\zeta)||d\theta$$
$$=\frac{1}{2\pi}\int_0^{2\pi}\log^+|f(\zeta)|d\theta+\frac{1}{2\pi}\int_0^{2\pi}\log^+\left|\frac{1}{f(\zeta)}\right|d\theta$$
$$=m(\rho,f)+m\left(\rho,\frac{1}{f}\right).$$

Therefore,

$$|I_1|\leqslant\frac{2\rho(p+1)!}{(\rho-r)^{p+2}}\left[m(\rho,f)+m\left(\rho,\frac{1}{f}\right)\right]$$
$$\leqslant\frac{2\rho^{p+2}(p+1)!}{(\rho-r)^{p}}\cdot\frac{2T(\rho,f)+0(1)}{\rho^{p+1}}.$$

It follows (by the assumption of p) that when $\rho\to\infty$, I_1 converges to 0 uniformly.

Next from $|a_\mu|\leqslant\rho$, $|z|\leqslant r$,

$$\left|\frac{\bar{a}_\mu}{a_\mu z-\rho^2}\right|=\frac{1}{\left|\frac{\rho^2}{a_\mu}-z\right|}\leqslant\frac{1}{\rho-r},$$

and hence

$$\left|\sum_{|a_\mu|\leqslant\rho}\left(\frac{\bar{a}_\mu}{a_\mu z-\rho^2}\right)^{p+1}\right|\leqslant\frac{n(\rho,0)}{(\rho-r)^{p+1}}.$$

Similarly, we have

$$\left|\sum_{|b_\nu|\leqslant\rho}\left(\frac{\bar{b}_\nu}{b_\nu z-\rho^2}\right)^{p+1}\right|\leqslant\frac{n(\rho,\infty)}{(\rho-r)^{p+1}}.$$

By $f(0)\neq 0$,

$$N(e\rho,0)=N(\rho,0)+\int_0^{e\rho}\frac{n(t,0)}{t}dt$$
$$\geqslant n(\rho,0)\int_\rho^{e\rho}\frac{dt}{t}=n(\rho,0).$$

Similarly we have

$$N(e\rho,\infty)\geqslant n(\rho,\infty).$$

Thus

$$|I_2|\leqslant p!\ \frac{1}{(\rho-r)^{p+1}}[n(\rho,0)+n(\rho,\infty)]$$

$$\leqslant p!\ \frac{\rho^{p+1}}{(\rho-r)^{p+1}}\cdot\frac{N(e\rho,0)+N(e\rho,\infty)}{\rho^{p+1}}$$

$$\leqslant p!\ \frac{(e\rho)^{p+1}}{(\rho-r)^{p+1}}\cdot\frac{2T(e\rho,f)+O(1)}{(e\rho)^{p+1}}.$$

As before, one can derive as above that when $\rho\to\infty$, I_2 tends to 0 uniformly. Consequently, from (1. 1. 45), as $\rho\to\infty$, it is uniformly that

$$\frac{d^{p+1}}{dz^{p+1}}\log f(z)=\lim_{\rho\to\infty}(-1)^p\cdot p!\cdot$$

$$\left[\sum_{|a_\mu|\leqslant\rho}\frac{1}{(z-a_\mu)^{p+1}}-\sum_{|b_\nu|\leqslant\rho}\frac{1}{(z-b_\nu)^{p+1}}\right]. \tag{1. 1. 46}$$

The right side of (1. 1. 46), as analysed above, is also convergent uniformly in any bounded domain containing neither zeros nor poles of $f(z)$. One can integrate both sides of the above equation from 0 to z, and noting that the right side can be integrated termwise. The p-th successive integration of $(-1)^p p!\ \dfrac{1}{(z-a)^{p+1}}$ is

$$\frac{1}{z-a}+\frac{1}{a}+\frac{z}{a^2}+\cdots+\frac{z^{p-1}}{a^p}.$$

The p+1-th integration becomes

$$\log\left(\frac{z-a}{-a}\right)+\frac{z}{a}+\frac{z^2}{2a^2}+\cdots+\frac{z^p}{pa^p}.$$

Accordingly from (1. 1. 46), we have

$$\log f(z)=\lim_{\rho\to\infty}\Big\{\sum_{|a_\mu|\leqslant\rho}\Big[\log\Big(1-\frac{z}{a_\mu}\Big)+\frac{z}{a_\mu}+\frac{z^2}{2a_\mu^2}+\cdots+\frac{z^p}{pa_\mu^p}\Big]$$

$$-\sum_{|b_\nu|\leqslant\rho}\Big[\log\Big(1-\frac{z}{b_\nu}\Big)+\frac{z}{b_\nu}+\frac{z^2}{2b_\nu^2}+\cdots+\frac{z^p}{pb_\nu^p}\Big]\Big\}+P(z),$$

where $P(z)$ denotes a suitable polynomial of degree no greater than p; it is produced through the successive $p+1$ integrations.

Hence

$$f(z)=e^{P(z)}\lim_{\rho\to\infty}\prod_{\substack{|a_\mu|\leqslant\rho\\|b_\nu|\leqslant\rho}}\left\{\frac{\Big(1-\frac{z}{a_\mu}\Big)e^{\frac{z}{a_\mu}+\cdots+\frac{z^p}{pa_\mu^p}}}{\Big(1-\frac{z}{b_\nu}\Big)e^{\frac{z}{b_\nu}+\cdots+\frac{z^p}{pb_\nu^p}}}\right\}.$$

If $z=0$ is a zero or pole of $f(z)$ of multiplicity k, then by considering $\dfrac{f(z)}{z^k}$ or $z^k f(z)$ and by the above argument, we have

$$f(z)=z^k e^{P(z)} \lim_{\rho\to\infty} \prod_{\substack{|a_\mu|\leqslant\rho \\ |b_v|\leqslant\rho}} \left\{ \frac{(1-\frac{z}{a_\mu})e^{\frac{z}{a_\mu}+\cdots+\frac{z^p}{pa_\mu^p}}}{(1-\frac{z}{b_v})e^{\frac{z}{b_v}+\cdots+\frac{z^p}{pb_v^p}}} \right\}, \tag{1.1.47}$$

where the right side is convergent uniformly on any bounded domain containing neither zeros nor poles of $f(z)$.

The numerator and denominator of the right side of (1. 1. 47) converge uniformly to infinite products independently, so, for any z, $f(z)$ can be represented as a quotient of two infinite products a necessary and sufficient condition is that both the numerator and denominator converge independently to finite values for all the values z.

Now if

$$\lim_{\rho\to\infty} \sum_{|a_\mu|\leqslant\rho} \frac{1}{|a_\mu|^{p+1}} \quad \text{and} \quad \lim_{\rho\to\infty} \sum_{|b_v|\leqslant\rho} \frac{1}{|b_v|^{p+1}}$$

converge to finite values, then, clearly, the numerator and denominator of the right side of (1. 1. 47) converge uniformly to finite values in any bounded domain. By Poisson-Jensen formula, if $\lim_{r\to\infty}\int_{r_0}^{r}\frac{T(t,f)}{t^{p+2}}dt$ is finite, then the above two infinite series converge to finite values, this also proves the later part of the theorem.

2. The Normal Family of Meromorphic Functions

2.1. THE CONCEPT OF NORMAL FAMILY

Definition. Let $F=\{f_a(z)\}_{a\in I}$ (I: index set) be a family of functions that are meromorphic in a domain D. F is said to be normal in D if every sequence $\{f_n(z)\}$ of functions belonging to that family has a subsequence that converges spherically uniformly in general sense inside D.

Here we say converges spherically means that on the w-plane that touches tangentially at $w=0$ of the Riemann sphere Σ of radius $1/2$, the values w of $w=f(z)$ have the property that for any given $\varepsilon>0$, there exists positive $n_0(\varepsilon,z)$ such that whenever $m>n>n_0$, the corresponding spherical distance of two points $f_m(z)$, $f_n(z)$ satisfying

$$d\,(f_m(z),f_n(z))<\varepsilon$$

$\{f_n(z)\}$ is said converges spherically uniformly meaning in general sense that if G is an arbitrary closed subdomain in D, then the previous mentioned quantity $n_0(\varepsilon,z)$ will only depend on E and ε, but independent of the z' s that belongs to G.

Definition. Let F be a family of functions that are meromorphic in D. F is said to be normal at a point $z \in D$ if it is normal in some neighborhood of z (i. e. , in an open disk centered at z with sufficiently small radius).

Theorem 1.2.1. *Let F be a family of functions that are meromorphic in a domain D. A necessary and sufficient condition for F to be normal in D is that F is normal at each point z in D.*

Proof. The necessity is obvious.

We now prove the sufficiency. According to the hypothesis, to each point z in D corresponding a open disk C_z such that F is normal in C_z. We can construct a closed disk W_z that lies inside C_z. Then any sequence of functions $\{f_n(z)\}$ from F , a subsequence $\{f_{n_k}(z)\}$ can be so chosen that it converges spherically uniformly in the general sense in W_z; noting that D can be covered by a countable subset $\{W_{z_k}\}$ of the family $\{W_z\}(z \in D)$.

Let $\{f_n(z)\}$ be an arbitrary sequence chosen from the family $\{f_\alpha(z)\}$, then there exists a subsequence $\{f_{1,n}(z)\}$ of $\{f_n(z)\}$ which converges spherically uniformly on W_{z_1}. Next from $\{f_{1,n}(z)\}$ a subsequence $\{f_{2,n}(z)\}$ can be chosen so that it converges spherically uniformly on W_{z_2}, And so on , in general, if a subsequence $\{f_{k-1,n}(z)\}$ has been selected, then from it a subsequence of $\{f_{k,n}(z)\}$ can be selected so that the subsequence of $\{f_{k,n}(z)\}$ converges spherically uniformly on W_{z_k}. Obviously, $\{f_{k,n}(z)\}$ converges spherically uniformly on $W_{z_1} \cup W_{z_2} \cup \cdots \cup W_{z_k}$. Now choose, from the family of sequences $\{f_{k,n}(z)\}(k=1,2,\cdots,n=1,2,\cdots)$ constructed above, the sequence $\{f_{n,n}(z)\}(n=1,2,\cdots)$. We are going to show that $\{f_{n,n}(z)\}$ converges spherically uniformly in the general sense on D. Let W be an arbitrary closed domain in D. Then a finite number of the W_z s will cover D; let k_0 denote the largest subindex appeared on the W_zs. Then clearly for $n \geqslant k_0$, the corresponding subsequence $\{f_{k_0,n}(z)\}$ is a subsequence of the sequence $\{f_{n,n}(z)\}$, therefore, it converges spherically uniformly on $W_{z_1} \cup W_{z_2} \cup \cdots \cup W_{z_{k_0}}$ and hence, on D.

2.2. THE PROPERTIES OF THE NORMAL FAMILIES OF MEROMORPHIC FUNCTIONS

Let $d(w_1, w_2)$ denote the arc (interior one on the great circle) length of two points

w'_1, w'_2 on the Riemann sphere, which are stereographic projections of the two planar point w_1, w_2 respectively. Now let $\bar{d}(w_1, w_2)$ denote the length of the chord $\overline{w_1' w'_2}$, then it is easy to verify that

$$\bar{d}(w_1, w_2) \leqslant d(w_1, w_2) \leqslant \frac{\pi}{2} \bar{d}(w_1, w_2).$$

Let

$$w'_1 = (\xi_1, \eta_1, \zeta_1), \quad w'_2 = (\xi_2, \eta_2, \zeta_2),$$

then

$$\begin{aligned} \bar{d}(w_1, w_2) &= \sqrt{(\xi_1-\xi_2)^2+(\eta_1-\eta_2)^2+(\zeta_1-\zeta_2)^2} \\ &= \frac{|w_1-w_2|}{\sqrt{1+|w_1|^2}\sqrt{1+|w_2|^2}} \\ &= \frac{\left|\frac{1}{w_1}-\frac{1}{w_2}\right|}{\sqrt{1+\frac{1}{|w_1|^2}}\sqrt{1+\frac{1}{|w_2|^2}}}. \end{aligned} \tag{1.2.1}$$

Define

$$\bar{d}(z, \infty) = \frac{2}{(1+|z|^2)^{1/2}},$$

and note

$$d(z, 0) = d(1/z, \infty).$$

It follows that when

$$|w_1|, |w_2| < A \tag{1.2.2}$$

or

$$\left|\frac{1}{w_1}\right|, \left|\frac{1}{w_2}\right| < A \tag{1.2.3}$$

then

$$\bar{d}(w_1, w_2) \geqslant \frac{1}{1+A^2} |w_1 - w_2|, \tag{1.2.4}$$

or

$$\bar{d}(w_1, w_2) \geqslant \frac{1}{1+A^2} \left|\frac{1}{w_1} - \frac{1}{w_2}\right|. \tag{1.2.5}$$

We note that when

$$\bar{d}(w_1, w_2) < 1 - \frac{2}{\sqrt{1+A^2}}, \quad A > 1 \tag{1.2.6}$$

holds then either (1.2.2) or (1.2.3) must be true. The reason is as follows: suppose that neither (1. 2. 2) nor (1. 2. 3) is true, we may assume without loss of generality that $|w_1| \geqslant A, \frac{1}{|w_2|} \geqslant A$. Then from

$$\bar{d}(\infty, 0) \leqslant \bar{d}(\infty, w_1) + \bar{d}(w_1, w_2) + \bar{d}(w_2, 0)$$

we have

$$\bar{d}(w_1, w_2) \geqslant \bar{d}(\infty, 0) - \bar{d}(\infty, w_1) - \bar{d}(w_2, 0)$$

$$= 1 - \frac{\frac{1}{|w_1|}}{\sqrt{1 + \frac{1}{|w_1|^2}}} - \frac{|w_2|}{\sqrt{1 + |w_2|^2}}$$

$$\geqslant 1 - \frac{\frac{1}{A}}{\sqrt{1 - \frac{1}{A^2}}} - \frac{\frac{1}{A}}{\sqrt{1 + \frac{1}{A^2}}} = 1 - \frac{2}{\sqrt{1 + A^2}}$$

This contradicts with (1. 2. 6).

From the above observation we see that the statement: sequence $\{f_n(z)\}$ converges spherically uniformly in a domain D is equivalent to say that either $|f_m(z) - f_n(z)| < \varepsilon$ or $\left|\frac{1}{f_m(z)} - \frac{1}{f_n(z)}\right| < \varepsilon$ must be held for $m > n > n(\varepsilon, w)$; w is an arbitrary closed subset of D, which also is equivalent to say that to any point $z_0 \in D$, there corresponds a neighborhood $N(z_0)$ so that either $\{f(z)\}$ or $\{\frac{1}{f_n(z)}\}$ converges uniformly in $N(z_0)$. It follows that if $\{f_n(z)\}$ converges spherically uniformly in a domain D, then its limit function is always meromorphic in D.

Theorem 1. 2. 2. *If the sequence of meromorphic functions $\{f_n(z)\}$ converges spherically uniformly to $f_0(z)$ on the closed domain W, then the sequence $\{\frac{\alpha f_n(z) + \beta}{\gamma f_n(z) + \delta}\}$ $(\alpha\delta - \beta\gamma \neq 0)$ also converges spherically uniformly to $\frac{\alpha f_0(z) + \beta}{\gamma f_0(z) + \delta}$.*

Proof.

$$\bar{d}\left(\frac{\alpha w_1 + \beta}{\gamma w_1 + \delta}, \frac{\alpha w_2 + \beta}{\gamma w_2 + \delta}\right)$$

$$= \frac{|\alpha\delta - \beta\gamma|\,|w_1 - w_2|}{\sqrt{|\alpha w_1 + \beta_1|^2 + |\gamma w_1 + \delta|^2}\sqrt{|\alpha w_2 + \beta|^2 + |\gamma w_2 + \delta|^2}}$$

$$= |\alpha\delta - \beta\gamma| \bar{d}(w_1, w_2). \sqrt{\frac{1+|w_1|^2}{|\alpha w_1+\beta|^2+|\gamma w_1+\delta|^2}} \sqrt{\frac{1+|w_2|^2}{|\alpha w_2+\beta|^2+|\gamma w_2+\delta|^2}}.$$

Now if $|\alpha\delta-\beta\gamma|\neq 0$, then to any w,

$$\frac{1+|w|^2}{|\alpha w+\beta|^2+|\gamma w+\delta|^2}<K<\infty.$$

The theorem follows.

Theorem 1. 2. 3. *If the sequence $\{f_n(z)\}$ of meromorphic functions converges spherically uniformly in the domain D in the general sense to function f_0, then f is meromorphic in D. Moreover, if none of the functions in the sequence $\{f_n(z)\}$ assumes the value a (finite or infinite) in D, then the limit function f_0 also does not assume the value a, unless f_0 is identically equal to a in D.*

Proof. The first half of the theorem has just been proved; the second half, according to Theorem 1. 2. 2, we only need to prove the case when $a=0$. We may assume that $f_0(z) \not\equiv 0$ and z be a zero of f. Then for a sufficiently small r_0, $\{f_n(z)\}$ converges uniformly to $f_0(z)$ on $|z-z_0|\leqslant 2r_0$. Furthermore, on $|z-z_0|\leqslant r_0$, $f_0(z)$ is regular and satisfies $|f_0(z)|<\frac{1}{2}A$ for some constant A. On the circle $|z-z_0|=r_0$, there exists a constant δ, such that $|f_0(z)|>2\delta$. Now choose n sufficiently large, then according to the hypothesis of the uniformly convergent, it follows that $|f_n(z)|<A$, and, hence, is regular on $|z-z_0|=r_0$. On $|z-z_0|\leqslant r$, $|f_n(z)|>\delta$, and by the hypothesis of the theorem, $f_n(z)$ has no zero on $|z-z_0|\leqslant 2r_0$; $\{f_n(z)\}$ converges uniformly to $f_0(z)$ on $|z-z_0|\leqslant 2r_0$. Therefore, $\{f'_n(z)\}$ converges uniformly to $f'_0(z)$ on $|z-z_0|=r_0$. Now since on $|z-z_0|=r_0$, $|f_n(z)|>\delta$, and $f'_n(z)/f_n(z)$ converges uniformly to $f'_0(z)/f_0(z)$, therefore for sufficiently large n_1, we have, for $n>n_1$

$$\left|\frac{f'_n(z)}{f_n(z)}-\frac{f'_0(z)}{f_0(z)}\right|<\frac{1}{r_0}$$

on $|z-z_0|=r$. It follows that

$$\left|\frac{1}{2\pi i}\int\frac{f'_n(z)}{f_n(z)}dz-\frac{1}{2\pi i}\int\frac{f'_0(z)}{f_0(z)}dz\right|<1 \qquad (n>n_1), \tag{1.2.7}$$

where the integration is along the positive direction of the circle. Viewing (1. 2. 7) in which the first term in the absolute value is the number of the zeros of $f_n(z)$ in $|z-z_0|<r$, which is zero, and the second term is the number of the zeros of $f_0(z)$ in $|z-z_0|<r$, which is >1. Hence the absolute value of the difference of these two terms is always

>1, which leads to a contradiction.

Corollary 1. *The limit function of a sequence of regular functions that converges spherically uniformly in the general sense in a domain D must be a regular function in D, unless it is identically equal to ∞.*

Corollary 2. *Let $\{f_n(z)\}$ be a sequence of meromorphic functions on a domain D. If $\{f_n(z)\}$ converges spherically uniformly to a function $f_0(z)$, which is not equal to constant a identically and never takes the value a on the boundary of a closed subdomain $W\subset D$. Then, for $n>n_0$, $f_n(z)$ and $f_0(z)$ have the same number of a-point in W.*

Proof. It follows immediately from (1.2.7).

Corollary 3. *Let $F=\{f(z)\}$ be a family of functions that are regular in the domain D. If any limit function of F is not identically equal to constant a, then, for an arbitrary closed subdomain $W\subset D$, the numbers of the a-points of functions in F are bounded above.*

Proof. If it is not bounded above, then a sequence of functions $\{f_n(z)\}\subset F$ can be chosen such that $f_n(z)$ has at least n a-points in W. Now choose from $\{f_n(z)\}$ a subsequence $\{f_{n_k}(z)\}$ which converges spherically uniformly in W, and, according to the hypothesis, its limit function cannot be a constant a. This contradicts with Corollary 2.

Consequently, we have the following conclusion.

Corollary 4. *Let $F=\{f(z)\}$ be a normal family of meromorphic functions in domain D and let E be a closed subdomain of D. Let z_0 be a point in W. If the value a is not a cluster point of the set $\{f(z_0)\}$, then, the numbers of the a-points of the functions in F that belong to are bounded above.*

Theorem 1.2.4. *Let $F=\{f_a(z)\}$ be a normal family of meromorphic functions in domain D. If a sequence $\{f_n(z)\}\subset F$ converges spherically uniformly in a subset W of D which has a limit point in D, then $\{f_n(z)\}$ converges spherically uniformly in the general sense in D.*

Proof. Since F is normal in D, it follows that a subsequence $\{f_{n_k}(z)\}$ of $\{f_n(z)\}$ can be chosen that converges spherically uniformly in D in the general sense. Let $f_0(z)$ be the limit function of $\{f_{n_k}(z)\}$, which, according to Theorem 1.2.3 is meromorphic in D. We now show it converges spherically uniformly in the general sense to $f_0(z)$ in D. Suppose it were not so. Then it would be the case for some closed subdomain $W \subset D$. It follows that there exist an infinite sequence of points $\{z_k\} \subset W$, a subsequence $\{f^*_{n_k}(z)\}$ of $\{f_n(z)\}$, and a positive number ε such that

$$d(f^*_{n_k}(z_k), f_0(z_k)) \geqslant \varepsilon. \tag{1.2.8}$$

Now on the other hand a subsequence $\{\tilde{f}_{n_k}(z)\}$ of $\{f^*_{n_k}(z)\}$ can be selected that will converge spherically uniformly in W. Let $f_0^*(z)$ be the limit function of $\{\tilde{f}_{n_k}(z)\}$, then by the hypothesis $f_0^*(z)$ and $f_0(z)$ coincide in W. Since $f_0(z)$ and $f_0^*(z)$ are meromorphic functions, it follows that $f_0(z) \equiv f_0^*(z)$ in D. This contradicts with (1.2.8).

Definition (Spherical Equicontinuity). Let $F = \{f_a(z)\}$ be a family of functions that are meromorphic in a closed domain W. If, to each given $\varepsilon > 0$, there exists a $\delta(\varepsilon) > 0$ such that for any two points z_1, z_2 in W satisfying $|z_1 - z_2| < \delta(\varepsilon)$, the following inequality

$$d(f(z_1), f(z_2)) < \varepsilon$$

holds for any function f in F, then F is called spherically equicontinuous in W.

Theorem 1.2.5. (Ascoli). *Let $F = \{f_n(z)\}$ be a sequence of meromorphic functions defined in a domain D that is spherically equicontinuous in a closed subset W of D. Then it is possible to select from F a subsequence that converges spherically uniformly in W.*

Proof. Choose a sequence of points $z_1, z_2, \cdots$, in W, which is also dense in W. Since for each point $z_i \in W$, the sequence $\{f_n(z_i)\}$ has at least one cluster point on the Riemann sphere Σ, therefore, from it a spherically convergent subsequence $\{f_{1,n}(z)\}$ can be chosen. Generally, if a sequence of meromorphic functions $\{f_{k-1,n}(z)\}$ has been defined, then a subsequence $\{f_{k,n}(z)\}$ can be selected, so that $\{f_{k,n}(z_k)\}$ is spherically convergent.

Now consider the sequence $\{f_{n,n}(z)\}$, it converges spherically at each point of the set $\{z_i\}$. We are going to show that actually the sequence $\{f_{n,n}(z)\}$ spherically uniformly converges on the entire set W.

Let $\varepsilon > 0$ be given. Let C_k denote the open disk of radius $\delta(=\varepsilon/3)$ with center at z_k. Then a finite many of such disks will cover W; let them be denoted $C_{a_1}, C_{a_2} \cdots, C_{a_p}$. Note that $\{f_{n,n}(z_{a_i})\}$ is convergent for each z_{a_i}. Hence when $m > n > N_{a_i}$, we have

$$d(f_{n,n}(z_{a_i}),\ f_{m,m}(z_{a_i}))<\frac{1}{3}\varepsilon. \tag{1.2.9}$$

Choose $N=\max_{1\leqslant i\leqslant p} N_{a_i}$. Now to any $z\in E$, it belongs to some C_{a_s}. Hence

$$|z-z_{a_s}|<\delta(\frac{\varepsilon}{3}).$$

It follows, from the hypothesis of spherical equicontinuity,

$$d(f_{n,n}(z),\ f_{n,n}(z_{a_s}))<\frac{1}{3}\varepsilon, \tag{1.2.10}$$

and

$$d(f_{m,m}(z),\ f_{m,m}(a_s))<\frac{1}{3}\varepsilon. \tag{1.2.11}$$

Combining (1.2.9), (1.2.10), and (1.2.11), we obtain, for $m>n>N$,

$$d(f_{n,n}(z), f_{m,m}(z))<\varepsilon,$$

where N depends on ε and W, but is independent of z. This shows that $\{f_{n,n}(z)\}$ is spherically uniformly convergent on W.

Theorem 1.2.6. *A necessary and sufficient condition for a family of meromorphic functions F to be normal in a domain D is that F is spherically equicontinuous on arbitrary closed subdomain of D.*

Proof. The sufficiency follows from Theorem 1.2.5. We only need to prove the necessity. Suppose that F is not spherically equicontinuous on a certain closed subdomain $W\subset D$. This assumption implies that there exist a positive number $\varepsilon>0$, two sequences of points $\{z_n\}$ and $\{z'_n\}$ in E, and a sequence of functions $\{f_n(z)\}$ in F such that

$$|z_n-z'_n|<\frac{1}{n},$$

but

$$d(f_n(z_n),\ f_n(z'_n))\geqslant\varepsilon. \tag{1.2.12}$$

Since $\{z_n\}\subset W$, it follows that a subsequence $\{z_{n_k}\}$ of $\{z_n\}$ will converge to a point $z_0\in W$, consequently the corresponding sequence $\{z'_{n_k}\}$ satisfies $z'_{n_k}\to z_0$. On the other hand, since F is normal, the sequence $\{f_{n_k}\}$ contains a subsequence $\{f^*_{n_k}\}$ which converges spherically uniformly to $f_0(z)$ on W. Now we rewrite $\{n_k'\}$ to be $\{n\}$, hence, we may assume, at the beginning, that

$$z_n\to z_0,\ z_n'\to z_0$$

and

$$f_n \to f_0$$

uniformly, hence, when $n > n_0(\varepsilon/3)$

$$d(f_n(z_n), f_0(z_n)) < \frac{1}{3}\varepsilon, \tag{1.2.13}$$

$$d(f_n(z'_n), f_0(z'_n)) < \frac{1}{3}\varepsilon. \tag{1.2.14}$$

From (1.2.12), (1.2.13), and (1.2.14), we obtain

$$d(f_0(z_n), f_0(z_n')) > \frac{1}{3}\varepsilon.$$

But $f_0(z)$ is meromorphic on the Riemann sphere, when

$$z_n \to z_0, \ z_n' \to z_0$$

we should have

$$d(f_0(z_n), f_0(z'_n)) \to 0.$$

This is a contradiction. Hence F must be spherically equicontinuous on closed subdomains of W.

Definition. The quantity

$$\lim_{\triangle z \to 0} \frac{d(f(z_0+\triangle z), f(z_0))}{|\triangle z|}$$

is called the spherical derivative of $f(z)$ at z, and is denoted by $Df(z)$.

When $f(z_0) \neq \infty$, it is easy to see that

$$Df(z_0) = \frac{|f'(z_0)|}{1+|f(z_0)|^2}. \tag{1.2.15}$$

When $f(z_0) = \infty$, we define

$$Df(z_0) = \lim_{z \to z_0} Df(z).$$

Let $L_a(z)$ represents a linear function corresponding to a rotation of the Riemann sphere. Then

$$d(w_1, w_2) = d\left(\frac{1}{w_1}, \frac{1}{w_2}\right) = d(L_a(w_1), L_a(w_2)),$$

it follows that

$$Df(z_0) = D\frac{1}{f(z_0)} = DL_a(f(z_0)).$$

Theorem 1.2.7. *If $f(z)$ is meromorphic on a closed domain W, then $Df(z)$ is continuous and bounded on W.*

Proof. Let us first divide E into a finite number of subsets W_i: $W_1, W_2, \cdots, W_k$, such that no two distinct zeros or poles of f will fall into the same subset. However, in each W_i, either $f(z)$ or $\frac{1}{f(z)}$ is regular, and hence is continuous and bounded there. Now from the observation $Df(z) = D(1/f(z))$, one concludes easily that $Df(z)$ is continuous and bounded on W.

Theorem 1.2.8. *Let $\{f_n(z)\}$ be a sequence of meromorphic functions that are defined on a closed domain W and converges spherically uniformly to a meromorphic function $f_0(z)$ on W. Then $\{Df_n(z)\}$ converges uniformly to $Df_0(z)$ on W.*

Proof. As in the previous proof, we divide W into W_i s of finite closed subdomains such that $f(z)$ or $\frac{1}{f(z)}$ is regular on each of the W_i s. If $f(z)$ is regular on W, then, for sufficiently large n, $f_n(z)$ is regular on W. Thus $\{f'_n(z)\}$ converges uniformly to $f'_0(z)$, and hence, $\frac{|f'_n(z)|}{1+|f_n(z)|^2}$ converges uniformly to $\frac{|f'_0(z)|}{1+|f_0(z)|^2}$. If $\frac{1}{f_0(z)}$ is regular, then $\frac{1}{f_n(z)}$ converges uniformly to $\frac{1}{f_0(z)}$. It follows that $D\frac{1}{f_n(z)}$ converges to $Df_0(z)$ and, hence, $Df_n(z)$ converges uniformly to $Df_0(z)$. As this holds true on each of the W_i s, therefore, $Df_n(z)$ converges uniformly to $Df_0(z)$ on entire W.

Theorem 1.2.9(Marty). *Let $F = \{f_\alpha(z)\}_{\alpha \in I}$ (I: index set) be a family of functions that are meromorphic in a domain D. Then a necessary and sufficient condition for F to be normal in D is that to any closed subdomain W of D, there exists a constant $M(W)$ (depending on W only) such that to any $f \in F$ the following condition holds:*

$$Df(z) < M(W), \quad z \in W.$$

Proof. The nesessity. If the stated condition does not hold, then, a sequence of functions $\{f_n(z)\}$ and points $\{z_n\}$ can be chosen from F and W respectively such that

$$Df_n(z_n) > n. \tag{1.2.16}$$

Now choose from $\{f_n(z)\}$ a subsequence $\{f_{n_k}(z)\}$, which converges spherically uniformly to $f_0(z)$ on W. Since $f_0(z)$ is meromorphic on W, it follows that

$$Df_0(z)<M_0;\ z\in W,$$

for some constant M_0.

On the other hand, because $Df_n(z)$ converges uniformly to $Df_0(z)$, we have, when $n>n_0$ with n_0 sufficiently large,

$$Df_n(z)<M_0+1.$$

This contradicts with (1.2.16).

The sufficiency. From the definition, we have

$$d(f(z_1),f(z_2))\leqslant\int_{z_1}^{z_2}Df(z)\,|dz|<|z_2-z_1|\cdot M(\overline{D}_0),$$

(the integration path is along the segment $[z_1,z_2]$).

This shows that $\{f_a(z)\}$ is spherically equicontinuous on W. Since W is an arbitrary closed subdomain of D, it follows from Theorem 1.2.5 that $\{f_a(z)\}$ is normal in D.

Theorem 1.2.10. *Let $F=\{f_a(z)\}$ be a family of functions that are regular and uniformly bounded in a domain D. Then F is normal in D.*

Proof. Let us choose an arbitrary point z_0 from D and draw a disk: $|z-z_0|\leqslant r_0$ completely contained in D. Then, to any $f\in F$, we have $|f(z)|<M$, for $z\in D$. Hence when $|z-z_0|\leqslant n_0$,

$$|f'(z)|=\left|\frac{1}{2\pi i}\int_{|\zeta-z_0|=2r_0}\frac{f(\zeta)}{(\zeta-z)^2}d\zeta\right|\leqslant\frac{M}{r_0}$$

It follows that whenever $|z-z_0|\leqslant r_0$,

$$Df(z)\leqslant\frac{M}{r_0}.$$

By virtue of Theorem 1.2.8, $\{f_a(z)\}$ is normal on $|z_0-z|\leqslant r_0$. Hence it follows from Theorem 1.1.1 that $\{f_a(z)\}$ is normal in D.

We note from the proof the above theorem that the conclusion of Theorem 1.2.10 remains valid if the condition" $\{f_a(z)\}$ is uniformly bounded in D" is replaced by" $\{f_a(z)\}$ is uniformly bounded in arbitrary closed subdomains of D".

Corollary 4. *Let $F=\{f_a(z)\}$ be a family of functions that are regular in a domain D. If F is uniformly bounded in closed subdomains of D, then F is normal in D.*

Corollary 5. *Let $F=\{f_a(z)\}$ be a family of functions that are meromorphic in a domain D. If to each $f\in F$ and $z\in D$,*

$$|f(z)-a|\geqslant m>0,$$

for some value a; namely F fails to cover a certain region (non-empty) on the Riemann sphere, then F is normal in D.

Proof. This follows immediately from Theorems 1. 2. 9 and 1. 2. 1.

Corollary 6. (Vitali). *Let $F=\{f_n(z)\}$ be a sequence of functions that are meromorphic in a domain D. Suppose that there exists a value b such that $\{\frac{1}{f_n(z)-b}\}$ is uniformly bounded. If $\{f_n(z)\}$ converges on a set of points of D that has a cluster point in D, then $\{f_n(z)\}$ converges spherically uniformly in any closed subdomain of D (or equivalently to say that it converges in the interior of D).*

Proof. This follows as a consequence of Theorems 1. 2. 4 and 1. 2. 10.

Theorem 1. 2. 11. *Let $F=\{f_a(z)\}$ be a family of functions that are regular in a domain D. If at any arbitrary point $z_0\in D$, $\{f_a(z_0)\}$ is bounded, say by a constant, then $\{f_a(z)\}$ is uniformly bounded on closed subdomains of D.*

Proof. Suppose that F is not uniformly bounded on some closed subdomain W, this means that there exist a sequence $\{f_n(z)\}$ of functions $f\in F$ and points $\{z_n\}\subset W$ such that

$$|f_n(z_n)|>n. \tag{1. 2. 17}$$

But $\{f_n(z)\}$ contains a subsequence $\{f_{n_k}(z)\}$ that converges spherically uniformly to a function $f_0(z)$ on W. Since $|f_{n_k}(z_0)|<M$, it follows from Theorem 1. 2. 2 that $f_0(z)$ is regular on W, and, hence, for all $z\in W$, $|f_0(z)|<K$, for some constant K. Consequently, for sufficiently large k_0, whenever $k>k_0$

$$|f_{n_k}(z)|<|f_0(z)|+1<K+1.$$

However, this will be incompatible with (1. 2. 7) by choosing $n>K+1$. The theorem is thus proved.

3. The Distance of a Family of Functions at a Point

Let $F=\{f_a(z)\}$ be a family of functions that are meromorphic in a domain D. Consider $sup\{d(f(z_1),d(f(z_2)): f\in F, z_1,z_2\in\{z:|z-z_0|<\delta\}\}$, which is denoted by

$$O_s(\{f\},z_0,\delta). \qquad (1.3.1)$$

Note that (1. 3. 1) decreases montonically as δ does. Hence when $\delta\to 0$, the limit of (1. 3. 1) exists, and the quantity

$$\lim_{\delta\to 0} O_s(\{f\},z_0,\delta\}=O_s(\{f\},z_0) \qquad (1.3.2)$$

is called the distance of F at point z.

Theorem 1. 3. 1. *A necessary and sufficient condition for a family of meromorphic functions F defined in a domain D that are spherically equicontinuous on subdomains of D is that the distance of F at an arbitrary point of D is zero.*

Proof. The necessity is obvious. We shall only prove the sufficiency. Let z_0 be an arbitrary point in D, then $O_s(\{f\},z_0)=0$. Now if F is not spherically equicontinuous on some subdomain W of D, it implies that there exists $\varepsilon>0$, and sequence of functions $\{f_n(z)\}$ and sequence of points $\{z_n\}$ can be chosen from F and W respectively such that

$$d(f_n(z_n),f_n(z_n'))\geq\varepsilon \quad (|z_n-z_n'|<\frac{1}{n}). \qquad (1.3.3)$$

A subsequence $\{z_{n_k}\}$ can be chosen from $\{z_n\}$ that converges to a point $z_0\in W$. This will yield $z'_{n_k}\to z_0$. Note the fact $O_s(\{f\},z_0)=0$; we have, for arbitrarily small $\delta>0$, to any $f\in F$ whenever $|z'-z_0|<\delta$, $|z''-z_0|<\delta$

$$d(f(z'), f(z''))<\varepsilon,$$

this contradicts with (1. 3. 3).

Corollary. *Let $F=\{f_a(z)\}$ be a family of functions that are meromorphic in a domain D. A necessary and sufficient condition for F to be normal in D is that the distance of F at each point of D is zero.*

Thus if a family of functions $F=\{f_a(z)\}$ that are meromorphic in a domain D is not normal in D, it implies from Theorem 1. 2. 1 that $\{f_a(z)\}$ fails to be normal at some point $z_0\in D$. This also means $O_s(\{f\},z_0)>0$, and such a point z_0 is called an abnormal point.

Furthermore, if the family $F=\{f_a(z)\}$ is not normal at a point z_0, then, according to the definition of normality, in any open disk C_δ: $\{|z-z_0|<\delta\}$ there exists a sequence of functions $S_\delta=\{f_n(z)\}$ from F such that none subsequence of $\{f_n(z)\}$ can be convergent spherically uniformly in C_δ.

However, generally speaking, the subsequence S_δ mentioned above is dependent on δ. But if S_δ is independent on δ, that means, in any open disk $|z-z_0|<\delta'$, any subsequence of S_δ will not be convergent uniformly, and such a point z_0 is called an O—point; an abnormal point, is generally called a J-point.

In the investigation of the normality of a family of functions in general, a J-point may not be an O-point; however, for family of meromorphic functions each J-point is also an O-point as can be shown by the theorem below.

Theorem 1.3.2. *Let $F=\{f_a(z)\}$ be a family of functions that are meromorphic in a domain D. If F fails to be normal at a point $z_0\in D$, then there must be some sequence of funtions $\{f_n(z)\}$ of F such that none subsequence of $\{f_n(z)\}$ will be convergent spherically uniformly in $|z-z_0|<\varepsilon$ for arbitrary $\delta>0$.*

Proof. Since $O_\delta(\{f\},z_0)=\omega>0$, a sequence of functions $\{f_n(z)\}$ and sequences of points $\{z_n\}$ and $\{z_n'\}$ can be chosen such that

$$|z_n-z_0|<\frac{1}{n},\quad |z_n'-z_0|<\frac{1}{n},$$

and

$$d(f_n(z_n),\ f_n(z'_n))>\frac{1}{2}\omega.$$

The fact that meromorphic functions are continuous spherically and the condition $z_n\to z_0$, $z'_n\to z_0$ indicates that none of the functions in $\{f_n(z)\}$ can occur infinitely many times. It follows that any subsequence of $\{f(z)\}$ will not be convergent spherically uniformly in $|z-z_0|\leqslant\delta$; δ an arbitrary positive number.

We now know that when $\{f_n(z)\}$ is not normal at point z_0, then $\{f_n(z)\}$ will not be spherically uniformly convergent in a neighborhood of z_0. However, we note that there may be a subsequence of $\{f_n(z)\}$ convergent spherically uniformly in a neighborhood of z_0. For instance, a function $f(z)$ that is regular in $|z|<1$ having at least one singularity on $|z|=1$ can be represented by a Taylor series:

$$f(z)=a_0+a_1z+\cdots+a_nz^n+\cdots,$$

which converges uniformly in the general sense (i. e. , in closed subsets of $|z|<1$). It follows the sequence

$$\{S_n(z)\}=\left\{\sum_{k=0}^{n}a_kz^k\right\}$$

of the partial sums of f converges uniformly in the general sense in $|z|<1$. But it is possible that there is some subsequence $\{S_{n_k}(z)\}$ of $\{S_n(z)\}$ satisfying suitable condition will be convergent at a neighborhood of a point on the circumference $|z|=1$, say $z=1$. When this phenomenon occurs it is called overconvergent.

Ostrowski' s theorem. *Let*

$$f(z)=\sum_{n=0}^{\infty}a^nz^n$$

be regular in $|z|<1$, *and, moreover, it is regular at a point* z_0 *on* $|z|=1$. *Let* $\varepsilon>0$ *be given and* $n_0<n_1<\cdots<n_k<\cdots$ *be a sequence of positive integers satisfying*

$$n_0=0,\ n_{2k}>(1+\varepsilon)n_{2k-1}\qquad (k=1,2,\cdots).$$

Suppose now that whenever n satisfies $n_{2k-1}<n<n_{2k}\ (k=1,2,\cdots)$ *then* $a_n=0$ *in the above expansion. Then*

$$\{S^k(z)\}=\left\{\sum_{n=0}^{n_{2k+1}}a_nz^n\right\}$$

converges uniformly in a sufficiently small neighborhood of z_0.

Proof. Choose integer $p>1/\varepsilon$, and let

$$z=\frac{1}{2}z_0(\zeta^p+\zeta^{p+1}),$$

$$f(z)=g(\zeta),$$

then determine $b_0,b_1,\cdots,b_m,\cdots$, so that the following equations hold:

$$\sum_{n=n_{2k}}^{n_{2k+1}}a_n\left[\frac{1}{2}z_0(\zeta^p+\zeta^{p+1})\right]^n=\sum_{m=pn_{2k}}^{pn_{2k+2}-1}b_m\zeta_m\qquad (k=0,1,2\cdots).$$

Thus

$$g(\zeta)=\sum_{k=0}^{\infty}\sum_{m=pn_{2k}}^{pn_{2k+2}-1}b_m\zeta^m\qquad (|\zeta^p+\zeta^{p+1}|<2),$$

$$\sum_{m=pn_{2k}}^{pn_{2k+2}-1}|b_m\zeta^m|\leqslant\sum_{n=n_{2k}}^{n_{2k+1}}\frac{|a_n|}{2^n}(|\zeta|^p+|\zeta|^{p+1})^n.$$

Noting that $\sum_{n=0}^{\infty}|a_n|z^n$ converges in $|z|<1$, we can conclude that

$$\sum_{k=0}^{\infty}\sum_{m=p^n_{2k}}^{p^n_{2k+2}-1}|b_m\zeta^m|$$

is convergent in $|z|<1$. It follows that

$$g(\zeta)=\sum_{m=0}^{\infty}b_m\zeta^m \qquad (|\zeta|<1).$$

On the other hand, we can show that $g(\zeta)$ is regular on $|z|\leqslant 1$. The reason is that under $|\zeta|\leqslant 1$, only $\zeta=1$ can lead to $|\zeta^p+\zeta^{p+1}|=2$, and hence $z=z_0$. If $\zeta\neq 1$, then $|z|<1$; namely $g(z)=\sum_{m=0}^{\infty}b_m\zeta^m$ is regular on $|z|\leqslant 1$, therefore it converges uniformly in $|\zeta|<r(r>1)$. Hence

$$g(\zeta)=\sum_{k=0}^{\infty}\sum_{m=p^n_{2k}}^{p^n_{2k+2}-1}b_m\zeta^m,$$

or $\{S_{n_{2k}}(z)\}$ converges uniformly in a region corresponding to $|\zeta|<r$ which is a neighborhood of z_0.

From the theorem one can immediately derive the following result.

Gap therorem (Hadamard). *Let*

$$f(z)=\sum_{n=0}^{\infty}a_{\lambda_n}z^{\lambda_n}$$

be a power series having $|z|<1$ as its circle of convergence. If $0\leqslant\lambda_1<\lambda_2<\cdots<\lambda_n<\cdots$, and $\varliminf_{n\to\infty}\frac{\lambda_{n+1}}{\lambda_n}>1$, then $|z|=1$ is the natural boundary of f.

Proof. By Ostrowski's theorem noting $(n_{2k}=n_{2k+1}=\lambda_k)$, if f is regular at a certain point on $|z|=1$, then $\{S_{\lambda_n}(z)\}=\left\{\sum_{n=1}^{k}a_{\lambda_n}z^{\lambda_n}\right\}$ will be convergent uniformly in a neighborhood of that point. However the convergence of $\{S_{\lambda_k}(z)\}$ is nothing but in the sense of the convergence of the Taylor series, which is impossible to be convergent at anywhere in $|z|>1$. Thus $f(z)$ cannot be regular at any point on $|z|=1$.

Theorem 1.3.3. *Let $\{f_a(z)\}$ be a family of functions that are meromorphic in a certain domain. Then at any of its abnormal point z_0, $O_s(\{f\},z_0)=\omega=\frac{\pi}{2}$.*

Note: Thus a family of meromorphic functions at any of its abnormal point has not only $\omega>0$ but also is impossible to have $0<\omega<\frac{\pi}{2}$, i. e. ω has to be equal to $\pi/2$.

Proof. Assume that $\omega<\pi/2$. Then choose $\eta>0$ such that $\omega+2\eta<\pi/2$. By the hypothesis, we have, for any $f\in\{f_\alpha(z)\}$ and sufficiently small δ,

$$d(f(z_0),f(z))<\omega+\eta \ (|z-z_0|<\delta).$$

Choose $\{f_n(z)\}$ a subsequence of $\{f_\alpha(z)\}$ such that $\{f_n(z_0)\}$ converges spherically to w_0. Then for $n>n_0(\eta)$

$$d(f_n(z_0),w_0)<\eta.$$

Hence under $|z-z_0|<\delta$, $n>n_0(\eta)$, we have $d(w_0,f_n(z))<\omega+2\eta$; namely $\{f_n(z)\}$ does not assume any value w that satisfies $d(w,w_0)>\omega+2\eta$. It follows from Theorem 1. 2. 10 and its corollary 2 that $\{f_n(z)\}$ and, hence, $\{f_\alpha(z)\}$ is normal at z, which is a contradiction. Thus we must conclude that $\omega=\pi/2$.

4. On Meromorphic Functions with Deficient Values

Theorem 1. 4. 1.(Schottky). *If $f(z)$ is regular and does not assume the values* 0 *and* 1 *in* $|z|<R$, *then in* $|z|<\theta R$

$$|f(z)|<\Omega(f(0),\theta). \tag{1. 4. 1}$$

Moreover, when

$$|f(0)|>\delta,\ |f(0)-1|>\delta,\ |\frac{1}{f(0)}|<\delta,$$

$$\Omega(f(0),\theta)\leqslant\Omega_1(\delta,\theta). \tag{1. 4. 2}$$

Where Ω,Ω_1 are constants that depend on the quantities appeared in the brackets corresponding only.

Proof. Let $w_1=0, w_2=1, w_3=\infty$, then

$$n(r,w_i)\equiv N(r,w_i)\equiv 0 \qquad (i=1,2,3);$$

$$T(r,f)=m(r,f).$$

By Theorem 1. 1. 13 we have

$$m(r,f)<D(\rho,r) \quad (r<\rho<R). \tag{1. 4. 3}$$

We first set

$$\delta_0=min\left(|f(0)|,|f(0)-1|,\left|\frac{1}{f(0)}\right|\right) \qquad (\delta_0\leqslant 1),$$

and consider $|f'(0)|\neq 0$ and note

$$C_{-4}=f(0),\ G(z)=f(z)(f(z)-1),$$

$$G_0=f(0)(f(0)-1),\ d=0,\ m_G=0(G_0\neq 0,\neq\infty),\ \beta_{-4}=f'(0).$$

By applying Theorem 1. 1. 13 and from (1. 1. 40), (1. 1. 24) we have, for $R>\rho>r>r_0$, with suitably fixed $r_0>0$,

$$\begin{aligned} m(r,f)&<D(\rho,r)<6\log^+ m(\rho,f)+2\log^+\frac{1}{\rho-r}\\ &+3\log^+ R+4\log^+\frac{1}{r_0}+6(\log^+\log^+ R+\log^+\log^+\frac{1}{r_0})\\ &+11\log\frac{1}{\delta_0}+16\log^+\log\frac{1}{\delta_0}+10\log 3+20\log 2+2k_1\\ &+\log\left|\frac{1}{f'(0)}\right|<6\log^+ m(\rho,f)+2\log^+\frac{1}{\rho-r}\\ &+K_1(\sigma_0,R,r_0)+\log\left|\frac{1}{f'(0)}\right|. \end{aligned}$$

After some calculations we can obtain the following relation:

$$R>r>\max(r_0,R-1),$$

$$\begin{aligned} m(r,f)&\leqslant M_0+2\{2+\log 2+2\log\frac{1}{R-r}\\ &+K_1(\delta_0,R,r_0)+\log\left|\frac{1}{f'(0)}\right|\}. \end{aligned} \tag{1. 4. 4}$$

Note in which M is a constant that satisfies

$$x-12\log^+ x\geqslant\frac{1}{2}x,\ x\geqslant M.$$

It follows that if $|f'(0)|\geqslant\frac{\delta_0}{2R}$, then

$$m(r,f)\leqslant M_0+2\{2+\log 2+2\log\frac{1}{R-r}+K_1(\delta_0,R,r_0)+\log\frac{2R}{\delta_0}\}. \tag{1. 4. 5}$$

However, if $|f'(0)|\leqslant\frac{\delta_0}{2R}$, then two cases will be considered separately:

1°. $|f'(z)|<\frac{\delta_0}{2R}$ in $|z|<R$. In this case, we have, for $|z|<R$

$$\begin{aligned} |f(z)|&=|f(0)+\int_0^z f'(z)dz|\\ &\leqslant|f(0)|+\int_0^z|f'(z)||dz|<\frac{1}{\delta_0}+\frac{1}{2}\delta_0<\frac{2}{\delta_0}, \end{aligned}$$

hence the theorem follows clearly.

2°. $|f'(z)|<\frac{\delta_0}{2R}$ in $|z|<A(A<R)$, but $|f'(a)|=\frac{\delta_0}{2R}$ for some point a on $|z|=A$. In this case we choose r satisfying $R>r>A$ and set

$$g(z)=f\left(\frac{r^2(z-a)}{az-r^2}\right), \tag{1.4.6}$$

$$g(0)=f(a)$$

then

$$|g(0)-f(0)|=|f(a)-f(0)|=\left|\int_0^a f'(z)dz\right|$$
$$\leqslant|a|\frac{\delta_0}{2R}<\frac{1}{2}\delta_0.$$

Consequently

$$|g(0)|\geqslant\frac{1}{2}\delta_0,\ |g(0)-1|\geqslant\frac{1}{2}\delta_0,$$

$$|g(0)|\leqslant\frac{1}{\delta_0}+\frac{\delta_0}{2}\leqslant\frac{2}{\delta_0},$$

$$|g'(0)|=|f'(a)|\frac{r^2-A^2}{r^2}=\frac{\delta_0}{2R}\,\frac{r^2-A^2}{r^2}, \tag{1.4.7}$$

moreover, $g(z)$ is regular and does not assume $0,1$ in $|z|<\frac{r^2(A+R)}{RA+r^2}$.

This becomes exactly the same case as (1.4.4) (by setting $R_1=\frac{r^2(A+R)}{RA+r^2}$), Hence, for $R_1>r>max(R_1-1,r_0)$

$$m(r,g)\leqslant M_0+2\{2+log2+2log\frac{1}{R_1-r}$$
$$+K_1(\frac{\delta_0}{2},R_1,r_0)+log\left|\frac{1}{g'(0)}\right|\}.$$

Now, since $R>R_1$, if $n>max(R-1,r_0)$, then

$$m(r,g)\leqslant M_0+\{2+log2+2log\frac{RA+r^2}{r(r-A)(R-r)}$$
$$+K_1(\frac{\delta_0}{2},R_1,r_0)+log\frac{2r^2R}{\delta_0(r^2-A^2)}\}.$$

But from (1.4.7), we have

$$m(r,f)\leqslant\frac{r+A}{r-A}m(r,g),$$

hence, for $r>max(\frac{R+A}{2},R-1,r_0)$, we obtain

$$m(r,f)\leqslant\frac{4R}{R-r}\{M_0+2[2+log2+2log\frac{3R}{(R-r)}$$

$$+K_1(\frac{\delta_0}{2},R_1,r_0)+\log\frac{R^2}{\delta_0(R-r)}]\}. \tag{1.4.8}$$

Noting that $m(r,f)$ is an increasing function of r, we have for

$$R>r>\max(A,R-1,r_0),$$

$$m(r,f)\leqslant m(\frac{R+r}{2},f)\leqslant\frac{8R}{R-r}\{M_0+2[2+\log 2+2\log\frac{12R}{(R-r)^2}$$

$$+K_1(\frac{\delta_0}{2},R,r_0)+\log\frac{2R^2}{\delta_0(R-r)}]\}.$$

Thus when $|z|\leqslant A$,

$$|f(z)|\leqslant|f(0)|+A\cdot\frac{\delta_0}{2R}\leqslant\frac{1}{\delta_0}+\frac{\delta_0}{2}.$$

Hence, for $r>\max(R-1,r_0)$, we have

$$m(r,f)<\log(\frac{1}{\delta_0}+\frac{\delta_0}{2})+\frac{8R}{R-r}\{M_0$$

$$+2[2+\log 2+2\log\frac{12R}{(R-r)^2}+K_1(\frac{\delta_0}{2},R_1,r_0)+\log\frac{2R^2}{\delta_0(R-r)}]\}, \tag{1.4.9}$$

and, moreover,

$$\log|f(z)|<\frac{r+|z|}{r-|z|}m(r,f) \qquad (r>|z|).$$

Therefore if we set $r=\frac{R+|z|}{2}$, then from (1.4.5) and (1.4.9), and for $|z|<r=\theta R$, we have

$$|f(z)|<\exp\left\{K_2(\frac{1}{R-r},R,\delta_0,r_0)\right\}<K_3(\theta,R,\delta_0,r_0). \tag{1.4.10}$$

Here we note that K_2 is bounded when $1/(R-r)$ does. The theorem follows immediately and consider $f(Rz)$ instead of $f(z)$ and by observing that K_3 is independent of R.

Corollary 1. *The family* $F=\{f_a(z)\}$ *of functions that are regular and do not assume the values* 0, 1 *in a domain* D *is normal in* D.

Proof. By virtue of Theorem 1.2.1, it suffices to show that F is normal at each point of D. Now choose an arbitrary point z_0 and sufficiently small r_0 such that $|z-z_0|<r_0\subset D$.

1. Let $\{f_n(z)\}$ be a sequence of functions from F. Suppose that the set $\{f_n(z_0)\}$ has a cluster point w_0, which differs from 0, 1, then there exists a suitable subsequence $\{f_{n_k}(z)\}$ of $\{f_n(z)\}$ such that in $|z-z_0|<r_0$,

$$|f_{n_k}(z_0)|>\delta_0,\ |f_{n_k}(z_0)-1|>\delta_0,\ \left|\frac{1}{f_{n_k}(z_0)}\right|>\delta_0.$$

Since $f_{n_k}(z)$ is regular and $\neq 0,1$, we have, for $|z-z_0|<\theta r_0$,

$$|f_{n_k}(z)|<\Omega_1(\delta_0,0).$$

Accordingly, we can choose from $\{f_{n_k}(z)\}$ a subsequence which converges uniformly in $|z-z_0|\leqslant\theta' r_0 (0<\theta'<\theta)$; this shows that $\{f_a(z)\}$ is normal at z_0.

2. Suppose that the sequence $\{f_n(z_0)\}$ has no cluster points other than $0,1$. This means that there exists a subsequence $\{f_{n_k}(z_0)\}$ converges to one of the values $0,1$ and ∞. We may assume without loss of generality that 1 is the value; the other cases can be obtained by simply considering $1-f(z)$ or $1-1/f(z)$; we recall here that linear transformations preserve spherically uniformly convergent property, and hence, the normality. Let

$$g_{n_k}(z)=\frac{\log f_{n_k}(z)+2\pi i}{4\pi i},$$

$$\log f_{n_k}(z)=u_0+iv_0 \qquad (-\pi<v_0\leqslant\pi),$$

then when $k\to\infty, g_{n_k}(z_0)\to\frac{1}{2}$. But $g_{n_k}(z)\neq 0,1,\infty$, as argued in the above, there exists a subsequence $\{g_{n_k}(z)\}$ which converges spherically uniformly to $g_0(z)$ in $|z-z_0|\leqslant\theta' r_0$ $(0<\theta'<\theta)$, we are going to show that $g_0(z)$ has to be equal to $1/2$ identically. By the hypothesis that $f_{n_k}\neq 1$, it follows that $\log f_{n_k}(z)\neq 0$, and hence $g_{n_k}(z)$ does not assume $1/2$. It follows by Theorem 1. 2. 3 that $g_0(z)=1/2$ identically. Now since $\{g_{n_k}(z)\}$ converges to a constant $1/2$, therefore

$$f_{n_k}(z)=e^{2\pi i(2g_{n_k}(z)-1)}=e^{4\pi i g_{n_k}(z)}$$

converges uniformly to constant 1. The theorem is thus proved.

Corollary 2. *A family of functions $F=\{f_a(z)\}$ that are meromorphic in a domain D is normal in D if there are three distinct values w_1, w_2 and w_3 (may include ∞) such that any function f in F omits the three values in D.*

Proof. This follows immediately from Corollary 1 of Theorems 1. 4. 1 and 1. 2. 2.

Note that from Corollary 2 above we deduce easily that if a family of functions $F=\{f_a(z)\}$ meromorphic in a domain D is abnormal at a point $z_0\in D$, then each function in the family F assumes every vaue, except possibly two exceptional values, infinitely many times in an arbitrary neighborhood of z_0 in D: $|z-z_0|<\delta(\delta>0)$.

Note also that Picard theorem follows immediately from Theorem 1. 4. 1. Suppose that $f(z)$ is an entire function and omits two finite values, say 0 and 1, then to any given R, we have

$$|f(z)|<\Omega(f(0),\theta) \qquad (|z|<\theta R).$$

Since R can be arbitrarily large, by Liouville's theorem $f(z)$ must be a constant.

We have emphasized to unify the methods in our presentations; for instance, we derive Theorem 1. 4. 1 from Theorem 1. 1. 13. In generally speaking Theorem 1. 4. 1 (Schottky theorem), and, hence, Theorem 1. 1. 13 can be proved most precisely by utilizing modular functions. It had been proved earlier by Landau, Bloch, and Valiron by using some elementary methods. Among them, the most simplified one is the so called Bloch-Valiron-Landau method. Although it bears no important connection to the present book, however, the proof has its own merits and has something to do with the conformal mapping, we would like to illustrate the proof as follows.

Theorem 1. 4. 2.(Bloch). *Let function $w=f(z)$ be regular on $|z|\leqslant 1$ with $f'(0)=1$. Then $f(z)$ assumes every value of an open disk of radius B; here B is a constant independent of $f(z)$.*

Proof. We may assume without any loss of generality that $f(0)=0$. Set

$$F(\theta)=\theta\cdot max|f'(z)|, |z|\leqslant 1-\theta.$$

Then $F(\theta)$ is continuous on the interval $0\leqslant\theta\leqslant 1$ with $F(0)=0, F(1)=1$. It follows that there exists some θ_0 with $0<\theta_0\leqslant 1$ such that

$$F(\theta_0)=1;\ F(\theta)<1, 0\leqslant\theta<\theta_0.$$

Since $\max\limits_{|z|\leqslant|1-\theta|}|f'(z)|$ is attained on $|z|=1-\theta$, there are some z_0 with $|z_0|=1-\theta_0$ such that

$$|f'(z_0)|=\frac{1}{\theta_0}.$$

Now from the fact that $|z-z_0|\leqslant\frac{\theta_0}{2}$ is contained in $|z|\leqslant 1-\frac{\theta_0}{2}$ entirely and

$$F(\frac{\theta_0}{2})=\frac{\theta_0}{2}\cdot\max_{|z|\leqslant 1-\frac{\theta_0}{2}}|f'(z)|<1,$$

we have

$$|f'(z)|<\frac{2}{\theta_0}; \qquad |z-z_0|\leqslant\frac{\theta_0}{2}.$$

Hence, for $|z-z_0|\leqslant\frac{1}{2}\theta_0$,

$$|f(z)-f(z_0)|=|\int_{z_0}^{z} f'(\xi)d\xi|<1.$$

Now we proceed to show that, in the disk: $|z-z_0|<\frac{\theta_0}{2}$ $f(z)$ assumes every value in the disk $|w-f(z_0)|<\frac{1}{16}$.

In fact, if $f(z)\neq w_0$ in $|z-z_0|<\frac{\theta_0}{2}$, then

$$h(z)=\sqrt{1-\frac{f(z)-f(z_0)}{w_0-f(z_0)}}=1-\frac{f'(z)}{2(w_0-f(z_0))}(z-z_0)+\cdots$$

is regular in $|z-z_0|<\frac{\theta_0}{2}$, moreover,

$$|h^2(z)|\leqslant 1+\frac{|f(z)-f(z_0)|}{|w_0-f(z_0)|}\leqslant 1+\frac{1}{|w_0-f(z_0)|}.$$

It follows that

$$1+\frac{|f'(z_0)|^2}{4|w_0-f(z_0)|^2}\cdot\frac{\theta_0^2}{4}\leqslant\frac{1}{2\pi}\int_0^{2\pi}|h^2(\xi)|d\varphi(\xi=z_0+\frac{\theta_0}{2}e^{i\varphi})$$
$$\leqslant 1+\frac{1}{|w_0-f(z_0)|}.$$

This yields $|w_0-f(z_0)|\geqslant\frac{1}{16}$. Thus we have not just proved the theorem but also can conclude the absolute constant B, at least, $\geqslant 1/16$.

As a matter of fact, the disk of radius B in Theorem 1. 4. 2 is a univalent image of a certain subdomain in $|z|<1$ under the mapping $f(z)$; it indicates particularly that $z=\varphi(w)$, the inverse function of $w=f(z)$, is regular in that disk. However, in order to prove the Schottky theorem, it suffices to show that $f(z)$ assumes every value of a certain disk of radius B in the w-plane.

By using some theorems about univalent functions in the later sections (Theorems 1. 6. 3 and 1. 6. 5), one can obtain Theorem 1. 4. 2 more precisely as follows.

As before, choose z_0 with $|z_0|=1-\theta_0$ such that $f'(z_0)=\frac{\varepsilon}{\theta_0}$, $|\varepsilon|=1$;

$$|f(z)-f(z_0)|<1,\qquad |z-z_0|\leqslant\frac{\theta_0}{2}.$$

Set

$$g(\xi)=\frac{2}{\varepsilon}\{f(z_0+\frac{\theta_0}{2}\xi)-f(z_0)\},$$

then $g(\xi)$ is regular in $|\xi|<1$ and satisfies $|g(\xi)|<2$, $g(0)=0, g'(0)=1$. According to Theorem 1. 6. 5, $g(\xi)$ is a univalent function in $|\xi|<\rho_2(=2-\sqrt{3})$. Set

$$h(\xi)=\frac{1}{\rho_2}g(\rho_2\xi),$$

then $h(0)=0$, $h'(0)=1$. Moreover, $h(\xi)$ is a univalent function in $|\xi|<1$, it follows (from Theorem 1. 6. 3 by letting $\rho_2(1)=\frac{1}{4}$) that the image of $|\xi|<1$ under the mapping $h(\xi)$ contains an open disk of radius $\rho_2(1)$. Therefore, the image of $|\xi|<1$, under $g(\xi)$ contains a univalent disk of radius $\varphi_2(1)$. It follows that the image of the disk $|z-z_0|<\frac{\theta_0}{2}$ under the mapping $f(z)$ contains a univalent disk of radius $\frac{\rho_2\varphi_2(1)}{2}$.

The least upper bound of the B' s under the above more precise form is called the Bloch constant.

We now apply Bloch theorem to prove Schottky theorem as follows.

It suffices to show the case that $R=1$.

If $f(z)$ is regular and omits $0,1$ in $|z|<1$, then

$$g(z)=\log\left\{\sqrt{\frac{\log f(z)}{2\pi i}}-\sqrt{\frac{\log f(z)}{2\pi i}-1}\right\} \tag{1. 4. 11}$$

is regular and omits the following set of values in $|z|<1$,

$$\pm u_{m,n}=\pm\log(\sqrt{m}+\sqrt{m-1})+\frac{n\pi i}{2},$$

$$m=1,2,\cdots;\qquad n=0,\pm1,\pm2,\cdots. \tag{1. 4. 12}$$

Here, when $z=0$, imaginary part v of the logarithm is chosen to satisfy $-\pi<v\leqslant\pi$ and the argument of the square root to be in the semi-open interval $(-\frac{\pi}{2},\frac{\pi}{2}]$.

Since $f(z)$ omits $0,1$ in $|z|<1$, $\log f(z)$ is regular and omits values $0,1,\infty$ in $|z|<1$. Hence, both functions $\frac{\log f(z)}{2\pi i}$ and $\frac{\log f(z)}{2\pi i}-1$ are regular and omit values 0, $2\pi i$ in $|z|<1$. Thus the function in the bracket $\{\ \}$ of (1. 4. 11) is regular in $|z|<1$. Clearly, this function does not assume the value zero there, and hence, $g(z)$ is regular and omits the set of values (1. 4. 12) in $|z|<1$. Now by solving (1. 4. 11), we obtain

$$f(z)=-e^{\frac{\pi i}{2}(e^{2g(z)}+e^{-2g(z)})}. \tag{1. 4. 13}$$

If

$$g(z)=\pm\log(\sqrt{m}+\sqrt{m-1})+\frac{n\pi i}{2},$$

then

$$f(z)=-e^{\pm\frac{\pi i}{2}[(\sqrt{m}+\sqrt{m-1})^2+(\sqrt{m}-\sqrt{m-1})^2]}$$
$$=-e^{\pm\pi i(2m-1)}=1.$$

On the other hand the set $Z=\{\pm u_{m,n}\}_{\substack{m=1\\n=1}}^{\infty}$ distributes on the entire w-plane; the distances of two adjacent points of Z on the real axis are

$$u_{m+1,0}-u_{m,0}=\log\frac{\sqrt{m+1}+\sqrt{m}}{\sqrt{m}+\sqrt{m-1}}$$

$$=\log\frac{1+\sqrt{1+\frac{1}{m}}}{1+\sqrt{1-\frac{1}{m}}}.$$

These set of values attains its maximum $\log(\sqrt{2}+1)$ at $m=1$. The set Z can be obtained by adding the set $\{\frac{\pm n\pi i}{2}\}_{n=1}^{\infty}$ to each point of Z that lies on the real axis. It follows that centered at an arbitrary point in the w-plane draws a circle of radius larger than $\frac{1}{2}|\log(\sqrt{2}+1)+\frac{\pi i}{2}|$ (particularly a circle of radius 1) that will contain some points from the set Z.

We calculate $g'(z)$ the derivative of $g(z)$. If for some θ such that $g'(\xi)\neq 0$ in $|\xi|<\theta$, then the function

$$h(z)=\frac{g(\xi+(1-\theta)z)}{(1-\theta)g'(\xi)}$$

is regular with $h'(0)=1$ and omits all the values $\frac{\pm u_{m,n}}{(1-\theta)|g'(\xi)|}$ in $|z|<1$, where $u_{m,n}\in E$. Thus any circle with a radius $\frac{1}{(1-\theta)|g'(\xi)|}$ in the w-plane will contain at least one such omitted value. It follows that

$$\frac{1}{(1-\theta)|g'(\xi)|}>B\geqslant\frac{1}{16},\quad |g'(\xi)|<\frac{16}{1-\theta}.$$

Consequently, in $|z|<\theta$,

$$|g(z)|<|g(0)|+\frac{16\theta}{1-\theta}\equiv K_4(f(0),\theta),$$

that shows $|g(z)|$ is bounded above by a quantity depending on $g(0)$ (hence on $f(0)$ and θ). Hence by (1.4.13) for $|z|<\theta$

$$|f(z)|<e^{\pi\cdot e^{2K_4(f(0),\theta)}}\equiv K_5(f(0),\theta).$$

This also is the so called Schottky theorem.

Note that from (1. 4. 13) we see that if $f(0)$ does not near 0 or ∞, then $|g(0)|$ does not near ∞. Hence if $\delta<|f(0)|<\frac{1}{\delta}$, then $K_5(f(0),\theta)<K_6(\delta,\theta)$. But note that in the meantime $1-f(z)$ will behave similarly to that of $f(z)$ and will not assume values $0,1,\infty$. We obtain $|f(0)|<\frac{1}{\delta}$, and hence $K_5(f(0),\theta)<\Omega_2(\delta,\theta)$. This is also the most general form of Schottky theorem.

Theorem 1. 4. 3.(Schottky). *If $f(z)$ is regular with $|f(0)|<M$ and omits values $0,1$ in $|z|<1$, then $|f(z)|<\Omega_2$ for $|z|<\theta R$,* here *Ω_2 is a constant depending only on M and θ.*

Theorem 1. 4. 4.(Landau). *Let $f(z)$ be regular with $f'(0)\neq 0$ and omits values $0,1$ in $|z|<R$. Then R is no greater than a quantity that depends only on $f(0)$ and $|f'(0)|$.*

Proof. Let $\theta=\frac{1}{2}$, then by virtue of Theorem 1. 4. 3, we have, for $|z|<R/2$,

$$|f(z)|<\Omega_2(f(0),\frac{1}{2})\equiv K(f(0)).$$

Also from

$$f'(0)=\frac{1}{2\pi i}\int_{|z|=\frac{R}{2}}\frac{f(\xi)}{\xi^2}d\xi,$$

we derive

$$|f'(0)|\leqslant\frac{2K(f(0))}{R},\quad R\leqslant\frac{2K(f(0))}{|f'(0)|}.$$

Corollary. *If $f(z)$ is regular and omits $0,1$ in $|z|<R$ with $f^{(n)}(0)\neq 0$, then R is bounded by a constant that depends only on $f(0)$, $f^{(n)}(0)$, and n.*

Proof. The proof will be exactly the same as in the proof of Theorem 1. 4. 4. From

$$f^{(n)}(0)=\frac{n!}{2\pi i}\int_{|z|=\frac{R}{2}}\frac{f(\xi)}{\xi^{n+1}}d\xi,$$

it follows that

$$|f^{(n)}(0)|\leqslant n!\ (\frac{2}{R})^n K(f(0)),$$

$$R \leqslant 2\sqrt[n]{\frac{n!\ (K(f(0))}{f^{(n)}(0)}}.$$

There are several generalizations of Theorems 1. 4. 3 and 1. 4. 4. We state some of the main results as follows.

Let

$$f(z) = a_0 + a_1 z + \cdots + a_n z^n + \cdots$$

be regular with $|a_i| \leqslant K_i (i=0,1,2,\cdots,n)$ and assume the values 0,1 no more than n,m ($n \leqslant m$) time, respectively, in $|z|<R$. Then in $|z|<R$,

1. $$|f(z)| \leqslant \Omega_{n,m}(K_0, K_1, \cdots, K_n, \theta, R).$$

Here $\Omega_{n,m}(K_0, K_1, \cdots, K_n, \theta, R)$ is a function depending only on K_i, θ and R.

2. (Bieberbach). *Let $f(z) = a_0 + a_1 z + \cdots + a_n z_n + a_{n+1} z^{n+1} + \cdots$ be regular with $|a_i| \leqslant K_i (i=0,1,2,3,\cdots,n)$, $\theta_{n+1} \neq 0$ and assume the values 0,1 no more than $n,m (n \leqslant m)$ time, respectively, in $|z|<R$. Then $R \leqslant \Omega^*_{n,m}(K_0, K_1, \cdots, K_n, a_{n+1})$. Here $\Omega^*_{n,m}(K_0, K_1, \cdots, K_n, a_{n+1})$ is a function depending only on K_i and a_{n+1}.*

Theorem 1. 4. 3 and Theorem 1. 4. 4 are obtained by setting $m=n=0$ in the above two results, respectively.

Furthermore, Corollary 2 of Theorem 1. 4. 1 can be generalized to derive some sufficient conditions of normality for family of meromorphic functions.

(Caratheodory, Bloch). *Let $F = \{f_a(z)\}$ be a family of functions that are meromorphic in a domain D and $w_1, w_2, \cdots\cdots w_q$ be a set of a distinct values (may inclue ∞). Assume that for every function $f \in F$ except finitely many of them, the multiplicities of the roots of the equations $f(z) = w_i (i=1,2,\cdots,q)$ in D are no less than m_i, where the m_i satisfies $\sum_{i=1}^{q} \frac{1}{m_i} < q-2$. Then F is normal in D.*

(Bloch). *Let $F = \{f_a(z)\}$ be a family of functions that are regular in a domain D. If, to each function $f \in F$, the radius of the circle of univalence of the image $f(D)$ is bounded by a constant independent of f, then F is normal in D.*

(Ahlfors). *Let $F = \{f_a(z)\}$ be a family of functions that are meromorphic in a domain D. If the pre-image sets $\Delta_i = f^{-1}(W_i) (i=1,2,3)$ of an arbitrary function $f \in F$ are all lying outside of D, for three mutually exclusive simply connected regions W_1, W_2 and W_3 on the Riemann sphere that lies tangientially above the w-plane, then F is normal in D.*

5. The Applications of the Theory of Normal Families

Theorem 1. 5. 1.(Picard). *If $f(z)$ is a nonconstant meromorphic function, then it can omit at most two values (possibly including ∞).*

Proof. Suppose that $f(z)$ omits three values: w_1, w_2, and w_3. Note if necessarily we can consider

$$g(z)=\frac{w_3-w_2}{w_3-w_1}\cdot\frac{f(z)-w_1}{f(z)-w_2}.$$

Thus we may assume, without any loss of generality, that w_1, w_2, and w_3 are $0, 1, \infty$ respectively. Set

$$f_n(z)\equiv f(2^n z).$$

We now consider the sequence of functions $\{f_n(z)\}$ in $|z|<1$, and note that each $f_n(z)$ is regular and omits values $0, 1$ in $|z|<1$. By virtue of Corollary 1 of Theorem 1. 4. 1, $\{f_n(z)\}$ is normal in $|z|<1$. Furthermore, since $f_n(0)=f(0)$, we have, by Theorem 1. 2. 11, that $\{f_n(z)\}$ is bounded in $|z|<\theta$ $(0<\theta<1)$. Hence, in $|z|<\theta$,

$$|f_n(z)|=|f(2^n z)|<M.$$

This shows that $f(z)$ is bounded in the entire finite (complex) plane. According to Liouville theorem $f(z)$ must be a constant, which is a contradiction.

Theorem 1. 5. 2.(Picard). *Let $f(z)$ be a nonconstant meromorphic function defined in a domain D. Then f assumes every value, with two possible exceptions at most, in any neighborhood of an isolated essential singularity point of f.*

In other word, if $f(z)$ is regular for $0<|z-z_0|<r_0$, and there exist three distinct values a_1, a_2, a_3, such that $f(z)\neq a_i$ for $|z-z_0|<r_0$, then z_0 is not an essential singularity. Also note, actually the theorem implies that, f takes each of those nonexceptional values infinitely many times.

Proof. We may suppose that the essential singularity point is $z_0=0$. Choose r_0 sufficiently small, so that $f(z)$ is meromorphic in $0<|z|<r_0$. We will proceed as in the proof of the previous theorem, that a contradiction will be resulted if $f(z)$ omits $0, 1, \infty$ in $0<|z|<r_0$. Set

$$f_n(z)=f(\frac{z}{2^n})$$

and consider $\{f_n(z)\}$ in $\Gamma_0: \frac{r_0}{2}<|z|<r_0$. By assumption, $f_n(z)$ is regular and omits $0, 1$ in Γ_0, hence $\{f_n(z)\}$ is normal in Γ_0. Now choose a point z_0 in Γ_0. If the sequence $\{f_n(z_0)\}$ has a finite cluster point, then we can choose a subsequence $\{f_{n_k}(z)\}$ from $\{f_n(z)\}$ such that $|f_{n_k}(z_0)|<M$ for all the k. By Theorem 1. 2. 11, we have, for sufficiently small δ,

$$|f_{n_k}(z)|=|f(\frac{z}{2^{n_k}})|<M_1,$$

for $\frac{r_0}{2}+\delta<|z|<r_0-\delta$. This implies that $f(z)$ is bounded in a sequence of ring-shaped domains that shrinks to the origin. Particularly, f is regular, and hence, is bounded in a domain formed by two such ring-shaped domains: $0<|z|<r_1 (r_1<r_0)$. This contradicts the fact that $z=0$ is an isolated essential singularity of f.

If $\{f_n(z_0)\}$ has no finite cluster point, then $\{\frac{1}{f_n(z_0)}\}$ has 0 as its cluster point. Thus the same contradiction follows by considering $\{\frac{1}{f_n(z)}\}$; noting that if $f(z)$ omits $0, 1, \infty$, so does $1/f(z)$.

Theorem 1. 5. 3 (Julia). *Let $f(z)$ be a transcendental entire function. Then there exists at least one direction: $Arg\, z=\varphi_0$ such that given any $\varepsilon>0$, $f(z)$ takes every value, with two possible exceptions at most, in the sector: $\varphi_0-\varepsilon<Arg\, z<\varphi_0+\varepsilon$, an infinity of times. $Arg\, z=\varphi_0$ is called a Julia direction (of f).*

Proof. First we show that $\{f_n(z)\}=\{f(2^n z)\}$ is not a normal family in $\frac{1}{4}<|z|<4$. According to Weierstrass theorem (or Picard theorem above), when $k\to\infty$, there exists $\{z_k\}$ with $|z|\to\infty$ satisfying $|f(z_k)|<1$ such that to each z_k corresponding an integer n_k satisfying $\frac{1}{2}<|\frac{z_k}{2^{n_k}}|<2$. It follows that $n_k\to\infty$ when $k\to\infty$. Now if $\{f_n(z)\}$ is normal, then a subsequence $\{f_{n'_k}(z)\}$ of $\{f_{n_k}(z)\}$ can be selected so that it converges uniformly to a limit function $f_0(z)$ on $|z|\leqslant 2$. Since

$$|f_{n'_k}(\frac{z_{n'_k}}{2^{n'_k}})|<1,$$

hence $f_0(z)$ cannot be a constant. Therefore, $\{f_{n'_k}(z)\}$ is bounded on $\frac{1}{2}\leqslant|z|\leqslant 2$. Consequently, $f(z)$ is bounded in $2^{n'_k-1}<|z|<2^{n'_k+1}$, hence, it is bounded in the entire finite plane. This is inconsistent with the assumption that f is a nonconstant function.

Thus $\{f_n(z)\}$ cannot be normal in $\frac{1}{4}<|z|<4$. Then by Theorem 1. 4. 1 there exists at least one abnormal point, say $z=z_0$ in $\frac{1}{4}<|z|<4$. Furthermore, according to the remark of Corollary 2 of the Theorem 1. 4. 1, for any arbitrary but fixed $\delta>0$, there are infinitely many functions $f_n(z)$ in the sequence $\{f_n(z)\}$ such that each of the functions takes every value, with two common possible exceptions at most, an infinity of times in $|z-z_0|<\delta$. It follows that f takes every value, with two possible exceptions at most, an infinity of times in $|z-2^n z_0|<\delta\cdot 2^n$. Since δ can be chosen arbitrarily small, the argument: $Arg\ z_0=\varphi_0$ is then a Julia direction of f.

Examples of Julia direction

1. $Arg\ z=\pm\frac{\pi}{2}$ are Julia directions for $f(z)=e^z$. Let $\varepsilon>0$ be given. Then e^z takes every value other than 0, an infinity of times in $\frac{\pi}{2}-\varepsilon<Arg\ z<\frac{\pi}{2}+\varepsilon$; to any a ($\neq 0$, $\neq\infty$), $z=\log a+2n\pi i$ (a chosen fixed branch of $\log a$) will fall inside $\frac{\pi}{2}-\varepsilon<Arg\ z<\frac{\pi}{2}+\varepsilon$, when n is sufficiently large.

The case $Arg\ z=-\frac{\pi}{2}$ can be discussed similarly. Moreover, there is no other Julia direction for e^z. This can be shown by the fact that, to any given $\varepsilon>0$, within $|Arg\ z|\leqslant\frac{\pi}{2}-\varepsilon$, $|Arg\ z-\pi|\leqslant\frac{\pi}{2}-\varepsilon$, $\lim_{z\to\infty} e^z=0,\infty$ respectively.

2. $Arg\ z=0,\pi$ are Julia directions for $\sin z$. Given $\varepsilon>0$, $\sin z$ takes every finite value an infinity of times in $|Arg\ z|<\varepsilon$. In general a meromorphic function $f(z)$ with a period ω has $Arg\ z=\omega$, $Arg\ \omega+\pi$ as its Julia directions. By Picard theorem $f(z)$ takes every value, with two possible exceptions at most, an infinity of times; moreover, if $f(z_0)=a$, then $f(z_0\pm n\omega)=a$. Hence, to any given $\varepsilon>0$, when n is sufficiently large, $z\pm n\omega$ will fall into $|Arg\ z-Arg\ \omega|<\varepsilon$, $|Arg\ z-Arg\ \omega-\pi|<\varepsilon$ respectively.

3. Every direction is a Julia direction for the Weierstrass elliptic function: $\mathscr{P}(z)$. Since, to any given φ and $\varepsilon>0$, the sector: $\varphi-\varepsilon<Arg\ z<\varphi+\varepsilon$ contains an infinity of period-parallelogram. In general, an elliptic function has all the directions as its Julia directions.

Polya showed: The Julia directions of any meromorphic function formed a closed subset of the interval: $[0,2\pi]$ and, conversely, given any closed subset S of $[0,2\pi]$, one can construct an entire function of infinite order having S as its set of Julia directions.

The set of Julia directions of an infinite order entire function, and the singularities of a function regular in $|z|<1$ are related by the following facts:

Let $f(z)=\sum_{n=0}^{\infty} a_n z^n$ be an entire function of infinite order and $F(z)=\sum_{n=0}^{\infty} b_n z^n$ be a Taylor series with radius of convergence equal to 1. Then

I. $F(z)$ has at least one singularity on $|z|=1$;

I'. $f(z)$ has at least one Julia direction;

II. (Vivanti-Diense). If $|Arg\ b_n| \leqslant \frac{\pi}{2}-\delta$, $\delta>0$, then $z=1$ is a singularity;

II'. (Polya). If $|Arg\ a_n| \leqslant \frac{\pi}{2}-\delta$, $\delta>0$, then the positive x-axis is a Julia direction;

III. (Hurwitz-Polya). By suitably choosing $\varepsilon_n=\pm 1$, one can construct $F_1(z)=\sum_{n=0}^{\infty} \varepsilon_n b_n z^n$, having $z=1$ as its natural boundary;

III'. By suitably choosing $\varepsilon_n=\pm 1$, one can make $f_1(z)=\sum_{n=0}^{\infty} \varepsilon_n a_n z^n$ having all the directions as its Julia directions;

IV. (Fabry). Let $N(n)$ denote the number of nonzero terms in the first n terms of the sequence: $b_0, b_1, b_2, \cdots$. If $\lim_{n\to\infty} \frac{N(n)}{n}=0$, then the set: $|z|=1$ becomes the natural boundary of $F(z)$;

IV'. (Polya). Let $N(n)$ denote the number of nonzero terms in the first n terms of the sequence: $a_0, a_1, a_2, \cdots$. If $\lim_{n\to\infty} \frac{N(n)}{n}=0$, then every direction is a Julia direction for $f(z)$.

Theorem 1.5.3 can also be rephrased as follows. Let $f(z)$ be a transcendental entire function. Then given $\varepsilon>0$, there exists a sequence of points $\{z_n\}$ with $z_n\to\infty$ such that in

$$|z-z_n|<\varepsilon|z_n| \tag{1.5.1}$$

$f(z)$ takes every value, with two possible exceptions at most, an infinity of times.

Condition (1.5.1) can also be replaced by assuming

$$|z-z_n|<\varepsilon_n|z_n|,\ \varepsilon_n\to 0 \qquad (1.5.2)$$

where $\{\varepsilon_n\}$ is a sequence properly chosen. Moreover, it is known (Valiron-Milloux) that the rate of decrease of $\{\varepsilon_n\}$ is related to the increasing property of $T(r,f)$.

The same conclusion holds for a function $f(z)$ which is regular in a neighborhood of an isolated essential singularity z_0, or f may not be regular but omits at least one value in a neighborhood of z_0. Generally speaking, if there exists a continuous curve L which tends to the isolated essential singularity z_0 such that $f(z)\to a$ as $z\to z_0$ along L, then Theorem 1.5.1 holds.

Theorem 1.5.4. *Let $z=0$ be an isolated essential singularity of the function $f(z)$, which is meromorphic in $0<|z|<4r_0$. If there exist a value a and a continuous curve L converging to 0 such that $f(z)\to a$ as $z\to 0$ along L, then $\{f_n(z)\}=\{f(\frac{z}{2^n})\}$ is not normal in $\frac{r_0}{4}<|z|<4r_0$, it follows that there exists at least one direction: Arg $z=\varphi_0$ such that, for any given $\varepsilon>0$, $f(z)$ takes every value an infinity of times, with two possible exceptions at most, in $-\varepsilon<$ Arg $z<\varphi_0+\varepsilon$.*

Proof. Since $z=0$ is an isolated essential singularity, it follows from Weierstrass theorem that there exists a sequence of points $\{z_n\}$ such that $d(f(z_k),a)\geqslant\delta>0$ as $z_n\to 0$. Each z_k corresponded to an integer n_k satisfying $\frac{r_0}{2}<|2^{n_k}z_k|<2r_0$. Set $2^{n_k}z_k=z_k^*$.

If $\{f_n(z)\}$ is normal in $\frac{r_0}{4}<|z|<4r_0$, then a subsequence $\{f_{n_p}(z)\}$ of $\{f_{n_k}(z)\}$ can be selected that will converge spherically uniformly on $\frac{r_0}{2}<|z|\leqslant 2r_0$ to some limit function $f_0(z)$. Since

$$f_{n_p}(z_p^*)=f\left(\frac{z_p^*}{2^{n_p}}\right)=f(z_p),$$

it follows that $d(f_{n_p}(z_p^*),a)\geqslant\delta$, and hence $f_0(z)$ cann't be a constant equal to a. Furthermore, to any r satisfying $\frac{r_0}{2}<r<2r_0$ and $p=1,2,\cdots$, there exist $\theta_p(0\leqslant\theta_p<2\pi)$ and the corresponding points $\zeta_p=\frac{re^{i\theta_p}}{2^{n_p}}$ on L, such that when $p\to\infty$,

$$f_{n_p}(re^{i\theta_p})=f(\zeta_p)\to a.$$

The θ_p's has at least one cluster point, say θ_0, and $f_{n_p}(z)$ converges spherically uniformly to $f_0(z)$ on $\frac{r_0}{2}\leqslant|z|\leqslant 2r_0$ with $f_0(re^{i\theta_0})=a$. Such a θ_p exists that corresponds to each r

with $\frac{r_0}{2}<r<2r_0$, hence it must be that $f_0(z)\equiv a$. This is a contradiction and, hence $\{f_n(z)\}$ is not normal.

Thus far, we have only considered families of functions $\{f_n(z)\}=\{f(2^n z)\}$ or $\{f(\frac{z}{2^n})\}$ with Γ lying in $\frac{R}{4}<|z|<4R$ (in the whole finite plane case) or $\frac{r_0}{4}<|z|<4r_0$ (in $0<|z|<4r_0$ case), respectively. However, the same argument works for families of functions $\{f_n(z)\}=\{f(\varphi_n(z))\}$ as long as under $\varphi_n (n=1,2,\cdots)$ the images: $\bigcup\varphi_n(\Gamma)$ can cover the finite plane or the domain $0<|z|<4r_0$. Particularly, we can consider the family of functions $\{f_n(z)\}=\{f(2^{\pm n}e^{i\theta_n}z)\}$. Also let $\sigma=\sigma(t)(0\leqslant t<1)$ be a parametric representation of a continuous curve that converges to $z=\infty$ or $z=0$ as $t\to 1$. By considering the family of functions: $\{f_t(z)\}=\{f(\sigma(t)z)\}$; $0\leqslant t<1$, we will derive a similar conclusion; here Γ can be any ring-shaped domain: $r_1<|z|<r_2(r_1<r_2)$. We now state the case when the singularity z_0 being equal to ∞ as follows.

Theorem 1.5.5. *Assume that $f(z)$ is regular on $|z|\geqslant R$, which has $z=\infty$ as an isolated essential singularity. Let $\sigma(t)$ be a continuous curve started from a point on $|z|=R$ and ended at $z=\infty$ as t moving from 0 to 1. Now we, in an arbitrary domain: $r_1<|z|<r_2(R<r_1)$ and given $\varepsilon>0$, choose a suitable point z_0, then in the domain: $|z-z_0\sigma(t)|<\varepsilon|\sigma(t)|$, $f(z)$ takes every value with possible two exceptions at most.*

A meromorphic function may fail to have a Julia direction, if it fails to have an asymptotic value; such a function is called Julia exceptional function.

Theorem 1.5.6. (Marty). *Let $f(z)$ be meromorphic in the finite plane. Then a necessary and sufficient condition for the family of functions: $\{f_n(z)\}=\{f(2^n z)\}$ to be normal in $\frac{1}{4}<|z|<4$ is that*

$$Df(z)=O(\frac{1}{|z|}).$$

Proof. According to Theorem 1.2.9 that the following condition: that there exists a constant $M(\delta)$ satisfying $Df_n(\zeta)=\frac{2^n|f'(2^n\zeta)|}{1+|f(2^n\zeta)|^2}<m(\delta)(\frac{1}{4-\delta}\leqslant|\zeta|\leqslant 4-\delta)$ is a necessary and sufficient one for $\{f_n(z)\}$ to be normal in $\frac{1}{4}<|z|<4$. It is easy to see now that the assertion of the theorem follows from this.

Also we can conclude easily that, from the above theorem and the definitions of $A(r,f)$ and $T(r,f)$, any Julia function $f(z)$ must satisfy

$$A(r,f)=O(\log r),\ T(r,f)=O((\log r)^2).$$

Ostrowski obtained a necessary and sufficient condition for a function f to be a Julia exceptional function in terms of the distributions of the zeros and poles of f. First of all, the above growth condition implies that $f(z)$ is a meromorphic function of zero order. Hence by Theorem 1. 2. 3 $f(z)$ can be expressed as

$$f(z)=z^m\frac{\Pi(1-\frac{z}{a_\mu})}{\Pi(1-\frac{z}{b_\nu})}.$$

Then a necessary and sufficient condition for such a function $f(z)$ to be a Julia exceptional function is the following set of conditions:

1°. there exists a constant K_1 such that $|n(r,\infty)-n(r,0)|<K_1$;

2°. there exist constants K_2 and K_3(independent of r) such that $n(2r,\infty)-n(r,0)<K_2$, and $n(2r,0)-n(r,0)<K_3$ respectively;

3°. there exist constants K_4 and K_5 such that the following inequalities hold for any p and q:

$$|a_p|^m\frac{\prod\limits_{|a_\mu|<|a_p|}|\frac{a_p}{a_\mu}|}{\prod\limits_{|b_\nu|<|a_p|}|\frac{a_p}{b_\nu}|}<K_4,\qquad |b_q|^{-m}\frac{\prod\limits_{|b_\nu|<|b_q|}|\frac{b_q}{b_\nu}|}{\prod\limits_{|a_\mu|<|b_q|}|\frac{b_q}{a_\mu}|}<K_5;$$

4°. there exists a positive ε, such that, for any λ,μ,

$$|\frac{a_\lambda}{b_\mu}-1|\geqslant\varepsilon>0.$$

For example, if $q>1$, then

$$f(z)=\frac{\prod\limits_{n=0}^{\infty}(1-\frac{z}{q^n})}{\prod\limits_{m=0}^{\infty}(1+\frac{z}{q^m})}$$

is a Julia exceptional function.

Similarly, for a given meromorphic function $f(z)$, a necessary and sufficient condition for the family of functions: $\{f_n(z)\}=\{f(z+a_n)\}$ with $\{a_n\}$ being any arbitrary sequence to be normal in the entire finite plane is that

$$\frac{|f'(z)|}{1+|f(z)|^2}<K.$$

Consequently, any function $f(z)$ having the property mentioned above must satisfy:

$$T(r,f)=O(r^2).$$

Functions that satisfy the above condition include elliptic functions, bilinear functions of the exponential function e^z, rational functions, and Julia exceptional functions, etc. Particularly, for function f, its associated normal family of function $\{f_n(z)\}=\{f(z+a_n)\}$ fails to have constant functions as the limit function for any arbitrary sequence $\{a_n\}$, one can show further that such f shares similar properties with elliptic functions and can be regarded as a generalization of an elliptic function.

Similar discussions can be applied to sectorial domains and the following results can be obtained.

Theorem 1.5.7. *Let $f(z)$ be regular in the domain $D: a<\text{Arg } z<\beta$ and omits the values 0, 1 in D. Assume that f is bounded on $\text{Arg } z=r$ for some $r(a<r<\beta)$. Then, for any given $\varepsilon>0$, $f(z)$ is bounded in $0<|z|<r_0$ with $a+\varepsilon\leqslant|\text{Arg } z|\leqslant\beta-\varepsilon$.*

Theorem 1.5.8. *Let $f(z)$ be regular and is either omiting the values 0, 1 or bounded in D: $a<\text{Arg } z<\beta$. Assume that on $\text{Arg } z=r$ for some $r(a<r<\beta)$, $f(z)\to\omega_0$ as $z\to 0$. Then $f(z)$ converges uniformly to ω_0 on $a+\varepsilon\leqslant\text{Arg } z\leqslant\beta-\varepsilon$, for any given $\varepsilon>0$.*

We note that the same conclusion holds if the straight line $\text{Arg } z=r$ is replaced by any continuous curve L that lies in $a+\varepsilon\leqslant\text{Arg } z\leqslant\beta-\varepsilon(\varepsilon>0$; arbitrary).

Theorem 1.5.9. *Let $f(z)$ be meromorphic in $a<\text{Arg} z<\beta$, $0<|z|<r_0$. Suppose that for some a', β' with $a<a'<\beta'<\beta$, the limits $\lim_{r\to 0} f(re^{ia'})=w_1$ and $\lim_{r\to 0} f(re^{i\beta'})=w_2$ exist. If $w_1\neq w_2$, then $f(z)$ has at least one Julia direction: $\text{Arg } z=r$ satisfying $a'\leqslant r\leqslant\beta'$.*

6. Application to Univalent Functions

In this section we consider only univalent functions regular in the disk $|z|<1$. We use σ to denote the set of univalent functions $f(z)$ regular in $|z|<1$ when $f(0)=0$ as well as $f'(0)=1$.

Theorem 1.6.1. *σ is normal in $|z|<1$.*

Proof. Since the arbitrary $f(z)$ belonging to σ is regular in $|z|\leqslant 1-\frac{\delta}{2}$ $(\delta>0)$, then the point z_0 given the minimal value of $|f(z)|$ on $|z|=1-\frac{\delta}{2}$ must satisty

$$|f(z_0)|\leqslant 1. \tag{1.6.1}$$

Moreover, as $\frac{f(z)}{z}$ is regular in $|z|\leqslant 1-\frac{\delta}{2}$ and cannot be zero, which takes the minimal value on $|z|=1-\frac{\delta}{2}$. But since $\frac{f(z)}{z}$ is equal to 1 at the origin, $|f(z_0)|\leqslant|z_0|<1$.

We consider $f^*(z)=\frac{f(z)}{f(z_0)}$, then $f^*(0)=0$, $f^*(z_0)=1$, $f^*(z)\neq 1(z\neq z_0)$, and $|f^*(z)|\geqslant 1$ on $|z|=1-\frac{\delta}{2}$.

We assume the branch of $g(z)=\sqrt{f^*(z)-1}$ such that $g(0)=i$, then $g(z)$ is regular in $|z|<1-\frac{\delta}{2}$. Let $W=f^*(z)$ and $W'=g(z)$, it follows that $W'^2+1=W$, and $|W|<1$ corresponds to $|W'^2+1|<1$, which is, in fact, the internal part of the lemniscate $|W'+i||W'-i|=1$. Since $W=f^*(z)$ assumes all values of $|W|<1$ in $|z|<1-\frac{\delta}{2}$, when $g(0)=i, W'=g(z)$ assumes all values of the lemniscate containing $W'=i$. Moreover, since $f^*(z)$ is univalent, $g(z)$ cannot assume those values of the lemniscate containing $W'=-i$. Selecting a sufficiently small positive number ε, then we have:

$$|g(z)+i|\geqslant\varepsilon>0$$

for all $g(z)$ and $|z|<1-\frac{\delta}{2}$. Hence, by the Corollary 2 of Theorem 1.2.10, we find that $\{g(z)\}$ is normal in $|z|<1-\frac{\delta}{2}$, and $g(0)=i$ is bounded for all the g's, then, according to Theorem 1.2.11, $\{g(z)\}$ is bounded in $|z|\geqslant 1-\delta$.

Hence, $f^*(z)=g(z)^2+1$, it follows that $f(z)=f(z_0)\,f^*(z)$ is also bounded in $|z|\leqslant 1-\delta$. Since δ is arbitrary, $\{f(z)\}$ is normal in $|z|<1$. From this proof, we can immediately obain the following theorem:

Theorem 1.6.2. *For all those functions $f(z)$ belonging to σ, they satisfy:*

$$|f(z)|\leqslant\varphi_1(r). \tag{1.6.2}$$

in $|z| \leqslant r < 1$.

Theorem 1. 6. 3. *For all those functions $f(z)$ belonging to σ, they satisfy:*

$$|f(z)| \geqslant \varphi_2(r). \tag{1.6.3}$$

on $|z| = r$.

Proof. If the inequality (1. 6. 3) doesn't hold, then there exist $\{f_n(z)\}$ and $\{\theta_n\}$ such that

$$|f_n(re^{i\theta_n})| < \frac{1}{n}.$$

Since $\{f_n(z)\}$ is normal in the domain $r-\delta<|z|<r+\delta$, where $f_n(z)\neq 0$ for $n=1,2,\cdots$, there is a subsequence $\{f_{n_k}(z)\}\subset\{f_n(z)\}$ such that $\{f_{n_k}(z)\}$ converges uniformly to $f_0(z)$, which must be a constant zero by Theorem 1. 2. 3. It follows that $\{f'_{n_k}(z)\}$ converges uniformly to $f'_0(z)$, i. e. a constant zero. This is a contradiction with $f'_{n_k}(0)=1$. Q. E. D.

This theorem shows that the image of $f(z)$ on domain $|z|\leqslant r$ contains the disk $|w|\leqslant\varphi_2(r)$.

Theorem 1. 6. 4. *If a sequence of univalent functions $\{f_n(z)\}$ in domain D converges spherically uniformly in general sense to the limiting function $f_0(z)\not\equiv$ constant, then $f_0(z)$ is univalent in D.*

Proof. If $f_0(z)$ is not univalent, then there exist two points z_1 and z_2 such that $f_0(z_1)=f_0(z_2)=\zeta_0$. Since $f_0(z)$ is not a constant, there exist two non-overlapping disks C_1 and C_2 with centers z_1 and z_2 respectively in D such that $f_0(z)\neq\zeta_0$ on the circles of C_1 and C_2. From the Corollary 2 of Theorem 1. 2. 3 we get that as n sufficiently large, $f_n(z)$ can take ζ_0 on C_1 and C_2. This is in contradiction with $f_n(z)$ to be univalent.

Corollary. *The limit of an arbitrary uniformly convergent sequence in σ is also in σ.*

Proof. Since σ contains only functions $f(z)$ that satisty $f(0)=0$ and $f'(0)=1$, any limiting function, therefore, can not be a contant. The corollary follows from this abviously.

In terms of the abstract sense, we say that σ is compact. This fact is useful in studying various properties of σ. For example, we have already known, by Theorem 1. 6. 2 that $\varphi(r)$ satisfies $|f(z)|\leqslant\varphi(r)$ in $|z|\leqslant r$, while by the corollary of Theorem 1. 6. 4, we find, in fact, that there exist functions $f(z)$ with $|f(z)|=\varphi(r)$, provided we have selected $\varphi(r)$ suitably.

Theorem 1. 6. 5. *Let $\{f(z)\}$ denote a family of functions that are regular in $|z|<1$ with $|f(z)|<M$, $f(0)=0$, and $f'(0)=1$. Then there exists a constant ρ_m such that any $f(z)$ belonging to $\{f(z)\}$ is univalent in $|z|<\rho_m$.*

Proof. Suppose that such a ρ_m does not exist, then we may select a subsequence $\{f_n(z)\}$ from $\{f(z)\}$, such that it satisfies:

$$f_n(z_n)=f_n(z'_n)(=w_n),\ |z_n|<\frac{1}{n},\ |z'_n|<\frac{1}{n}.$$

Since $\{f(z)\}$ is normal in $|z|<1$, we may select, according to Theorem 1. 2. 10, a subsequence $\{f_{n_k}(z)\}$ that converges uniformly to the regular function $f_0(z)$. Obviously, $f_0(z)$ satisfies $f_0(0)=0$ and $f'_0(0)=1$. If we assume suitably $\rho>0$, then we have $|f_0(z)|\geqslant\delta>0$ on $|z|=\rho$. Hence, when $0<|z|\leqslant\rho$, $|f_0(z)|\neq0$ holds. It is also obvious that $w_{n_k}\to0$. Thus we obtain:

$$\frac{1}{2\pi i}\int_{|z|=\rho}\frac{f'_0(z)}{f_0(z)}dz=1.$$

However, when $n>\frac{1}{\rho}$, it follows that:

$$\frac{1}{2\pi i}\int_{|z|=\rho}\frac{f'_n(z)}{f_n(z)-w_n}dz\geqslant2.$$

Thus $\{f'_{n_k}(z)\}$ and $\{f_{n_k}(z)-w_{n_k}\}$ converge uniformly to $f_0'(z)$ and $f_0(z)$ on $|z|=\rho$, respectively, which, however is impossible. This also completes the proof of the theorem.

Note that we may take $\rho_M=M-\sqrt{M^2-1}$ in the theorem, and there exist functions satisfying the condition of Theorem 1. 6. 5 and assuming some value twice on $|z|=M-\sqrt{M^2-1}$. Also the function $f(z)$ of the Theorem is not only univalent in $|z|<\rho_m$, but it is also starlike. By f being starlike in a domain D, it means $f(z)$ is univalent in D and if the image $f(z)$ of D contains a point w then it also contains all the points tw $(0\leqslant t\leqslant1)$.

We may also prove the existence of r_s, suth that any $f\in\sigma$ is starlike in $|z|<r_s$. The

least upper bound R_s of such a r_s is called the starlike constant.

Furthermore, we may prove the existence of r_k, such that any $f\in\sigma$ is convex in $|z|<r_k$. By $f(z)$ being convex in domain D is meant $f(z)$ is univalent in D, and if the image $f(z)$ of D contains points w_1 and w_2, then it contains also all the points $w_1+t(w_2-w_1)$ $(0\leqslant t\leqslant 1)$. The least upper bound R_k of such a r_k is called a convex constant.

The subset σ' derived from the functions of σ added with other extra conditions yields different kinds of constants. Particularly, we consider those functions $f(z)$ of σ having the form:

$$f(z)=z+a_1z+a_2z^2+\cdots+a_nz^n+\cdots.$$

Let n be fixed and φ_n be the least upper bound of all such a $|a_n|$, then

$$\varphi_n<n\left(1+\frac{1}{n-1}\right)^{n-1}.$$

Generally, a set of functions with a particular property may have universal constants that are unrelated to individual function examples of such constants are exhibited above.

Universal constants include more than that. It is one of the important objects in the studying of function theory, and the Bloch constant of Theorem 1. 4. 2 derived in this book is an example. Also we note that in Theorems 1. 4. 3 and 1. 4. 4, θ has not been fixed, and the term $f(0)$ is not a constant, this, however, should be included in the same scope of idealogy.

Next a simple fact about the theory of univalent function is that we may derive from it a theory regarding the set of functions, with each its member assuming each value no more than a fixed times say p. Such a theory is the so called the theory of p-valent functions, with its kinds of theorems already known. Hence, for the set of p-valent functions, substantial results may be derived from the universal constants that correspond to given properties of the functions.

As consequences, many theorems of the univalent functions may be derived easily by means of the idea of normality.

Though using the method of normal family may prove the existence of various kinds of universal constants, we may, however, find it difficult to derive these values precisely. But in the univalent function theory, it is really of important that we could determine them precisely.

We shall stop here for the discussion of the cases of univalent functions. In the following, some theorems related to univalence will be described.

Theorem 1. 6. 6.(Landau). *Let $f(z)$ be an arbitrary function regular in $|z|\leqslant 1$ with $f(0)=0$ and $f'(0)=1$. Also the image set of $|z|\leqslant 1$ under f is a closed domain D in the w-plane. Then the maximal value r_f of all these values r such that the entirely circumference $|w|=r$ lies in D is $\geqslant \rho$, where ρ is a positive constant independent of $f(z)$.*

Proof. If there does not exist such a ρ as stated in the theorem, then to each n, there exists a function $f_n(z)$ with $f_n(0)=0$ and $f'_n(0)=1$ and such that the interior of the image set of $|z|\leqslant 1$ under f_n does not intersect either circle: $|w|=1$ or circle: $|w|=\frac{1}{2^n}$. Thus for $|z|\leqslant 1$,

$$f_n(z)\neq \alpha_n, \neq \beta_n; \quad |\alpha_n|=1, \quad |\beta_n|=\frac{1}{2^n}.$$

Set

$$g_n(z)=\frac{\alpha_n-\beta_n}{f_n(z)-\beta_n},$$

then $g_n(z)$ does not assume $0, 1$ and ∞ in $|z|\leqslant 1$. By Corollary 1 of Theorem 1. 4. 1, we find that $\{g_n(z)\}$ is normal in $|z|<1$. However, as $n\to\infty$,

$$g_n(0)=\frac{\alpha_n-\beta_n}{-\beta_n}\to\infty,$$

thus $\{g_n(z)\}$ has a spherically uniformly convergent subsequence $\{g_{n_k}(z)\}$, with its limiting function being constant ∞.

From

$$f_n(z)=\beta_n+\frac{\alpha_n-\beta_n}{g_n(z)},$$

we find that $\{f_{n_k}(z)\}$ must be spherically uniformly cnvergent to constant 0. But $f'_{n_k}(0)=1$, which is a contradiction. This also completes the proof.

The assumption $f'(0)=1$ has been used to conclude that any uniformly convergent sequence in $\{f(z)\}$ cannot have a constant as its limiting function. Thus the above theorem remains valid for any set of functions that do not have any constant as a limiting function.

Theorem 1. 6. 7.(Valiron). *Suppose that $\{f(z)\}$ is a family of functions which are regular in* |

$z|\leqslant 1$ with $f(0)=0$, and that they have no constants as their limiting functions. Then for any given $\varepsilon>0$, there exists a $p(\varepsilon)>0$ such that any $f(z)$ in $\{f(z)\}$ takes all values in $|w|<p(\varepsilon)$, with the exception of those values in a sector with vertex at $w=0$, radius $p(\varepsilon)$ and vertex angle 2ε in the w-plane.

Proof. If such a $p(\varepsilon)$ does not exist, we may then select a subsequence $\{f_n(z)\}$ from $\{f(z)\}$, such that $f_n(z)\neq\alpha_n$, $\neq\beta_n$, as $|z|<1$, where $|\alpha_n|, |\beta_n|<\frac{1}{n}$ and

$$\varepsilon<\varphi_n=\operatorname{Arg}\beta_n-\operatorname{Arg}\alpha_n<2\pi-\varepsilon.$$

If $\frac{\pi}{2}\leqslant\varphi_n\leqslant\frac{3}{2}\pi$, then from $|\alpha_n-\beta_n|\geqslant|\beta_n|$, we have

$$\left|\frac{\beta_n}{\alpha_n-\beta_n}\right|\leqslant 1\leqslant\frac{1}{\sin\varepsilon}.$$

If $\varphi_n<\frac{\pi}{2}$ or $\varphi_n>\frac{3}{2}\pi$, then

$$|\beta_n|\sin\varepsilon<|\beta_n||\sin\varphi_n|<|\alpha_n-\beta_n|.$$

Hence

$$\left|\frac{\beta_n}{\alpha_n-\beta_n}\right|<\frac{1}{\sin\varepsilon}.$$

In any cases, we have

$$\left|\frac{\beta_n}{\alpha_n-\beta_n}\right|\leqslant\frac{1}{\sin\varepsilon}.$$

Let

$$g_n(z)=\frac{f_n(z)-\beta_n}{\alpha_n-\beta_n},$$

then $g_n(z)$ does not assume $0,1$ and ∞ in $|z|<1$. Hence, $\{g_n(z)\}$ is normal in $|z|<1$. But

$$|g_n(0)|=\frac{|\beta_n|}{|\alpha_n-\beta_n|}<\frac{1}{\sin\varepsilon}$$

is bounded. From Theorem 1.2.11, $\{g_n\}$ are uniformly bounded in $|z|\leqslant r\,(r<1)$. And thus:

$$f_n(z)=\beta_n+(\alpha_n-\beta_n)g_n(z)$$

is convergent uniformly to 0 in $|z|\leqslant r\,(r<1)$, which leads to a contradiction. This completes the proof.

Note that this theorem contains Theorem 1.4.4. Let

$$f_n(z)=a_0+a_1z+\cdots+a_nz^n+\cdots,\ a_1\neq 0$$

be regular in $|z|<R$ and does not take values 0 and 1, then

$$g(z)=\frac{f(Rz)-a_0}{Ra_1}=z+\cdots$$

is regular in $|z|<1$ and does not assume $\alpha=\frac{a_0}{Ra_1}$ and $\beta=\frac{1-a_0}{Ra_1}$. Therefore, for any arbitrarily given $\varepsilon>0$, it follows that:

$$|\alpha|>p(\varepsilon),\ |\beta|>p(\varepsilon),\ \angle\alpha o\beta<2\varepsilon,$$

with at least one of them valid.

Let $\angle\alpha o\beta=\theta$, then

$$\cos\theta=\frac{|\alpha^2|+|\beta^2|-|\alpha-\beta|^2}{2|\alpha||\beta|}=\frac{|a_0|^2+|1-a_0|^2-1}{2|a_0||-a_0|},$$

which depends only on a_0. Let ε be smaller than a particular quantity $q(a_0)$ that is merely related to a_0, then we have:

$$|\alpha|=\left|\frac{a_0}{Ra_1}\right|>p(q(a_0)),$$

$$|\beta|=\left|\frac{1-a_0}{Ra_1}\right|>p(q(a_0))$$

implying that:

$$R|a_1|<\frac{|a_0|}{p(q(a_0))},\ R|a_1|<\frac{|1-a_0|}{p(q(a_0))}.$$

CHAPTER 2

H^p Space

1. Harmonic and Subharmonic Functions

1.1. HARMONIC FUNCTIONS

Let $f(z)$ be a function regular in the unit disk: $|z|<1$. Define

$$M_p(r,f)=\begin{cases}\left\{\dfrac{1}{2\pi}\displaystyle\int_0^{2\pi}|f(re^{i\theta})|^p d\theta\right\}^{1/p}, & 0<p<\infty;\\ M(r,f)=\max\limits_{0\leqslant\theta\leqslant 2\pi}|f(re^{i\theta})|. \end{cases}$$

Definitions. The class of all the functions $f(z)$ that is regular in the unit disk satisfying $\lim\limits_{r\to 1} M_p(r,f)<\infty$ is denoted by H^p. The class of all the bounded regular functions defined in the unit disk is denoted by H^∞. The class of the real-valued functions $u(z)$ that is harmonic and satisfying $M_p(r,u)<\infty$ in the unit disk is denoted by $h^p(0<p\leqslant\infty)$.

Since

$$a^p\leqslant(a+b)^p\leqslant 2^p(a^p+b^p),\ a\geqslant 0,\ b\geqslant 0 \qquad (0<p<\infty),$$

it follows that a necessary and sufficient condition for a regular function $f\in H^\infty$ is that both the real and imaginary parts of f belong to h^p.

Obviously, both H^p and h^p are linear spaces. Moreover, when $0<p<q\leqslant\infty$, we have

$$H^q\subset H^p \text{ and } h^q\subset h^p.$$

Let

$$P(r,\theta)=\frac{1-r^2}{1-2r\cos\theta+r^2}$$

be the Poisson kernel. Then the following Poisson integral

$$u(z)=u(re^{i\theta})=\frac{1}{2\pi}\int_0^{2\pi}P(r,\theta-t)u(e^{it})dt \tag{2.1.1}$$

is used to indicate the relationship between the harmonic function and its boundary values within the unit disk.

In (2.1.1) if $u(e^{it})$ is replaced by an arbitrary continuous function $\psi(t)$ with $\psi(0)=\psi(2\pi)$, then $u(z)$ remains to be a function harmonic in $|z|<1$ and continuous in $|z|\leqslant 1$, with its boundary values satisfying $u(e^{it})=\psi(t)$.

In general, a function $u(z)$ represented by the following so called Poisson—Stieltjes integral:

$$u(z)=u(re^{i\theta})=\frac{1}{2\pi}\int_0^{2\pi}P(r,\theta-t)d\mu(t), \tag{2.1.2}$$

where $\mu(t)$ is of bounded variation on $[0,2\pi]$, is a harmonic function in $|z|<1$.

Lemma (Helly′ s selection principle). *It is possible to select from a uniformly bounded and uniformly bounded variation sequence of functions $\{\mu_n(t)\}$ on the interval $[a,b]$ a subsequence $\{\mu_{n_k}(t)\}$ which is convergent uniformly almost everywhere to a function $\mu(t)$ of bounded variation with respect to t on $[a,b]$. Furthermore, to any continuous function $\psi(t)$ defined on $[a,b]$, then we have the following relationship among Poisson-Stieltjes integrals:*

$$\lim_{k\to\infty}\int_a^b\psi(t)d\mu_{n_k}(t)=\int_a^b\psi(t)d\mu(t).$$

Proof. We refer the reader to Chap. 8 of Natason′ s book (Theory of Real Variable[29]).

Theorem 2.1.1. *In the unit disk the following three classes of functions are equivalent to each other:*

i. *The Poisson-Stieltjes integrals;*

ii. *The difference of two nonnegative harmonic functions;*

iii. h^1.

Proof. (i)⇒(ii) is obvious. Since each function of bounded variation $\mu(t)$ can be expressed as the difference of two bounded and nondecreasing functions, it follows that every Poisson-Stieltjes integral is a difference of two nonnegative harmonic functions.

(ii)⇒(iii). Suppose $u(z)=u_1(z)-u_2(z)$, where $u_1(z)$ and $u_2(z)$ are two nonnegative harmonic functions, then

$$\int_0^{2\pi} |u(re^{i\theta})|d\theta \leqslant \int_0^{2\pi} u_1(re^{i\theta})d\theta + \int_0^{2\pi} u_2(re^{i\theta})d\theta = 2\pi[u_1(0)+u_2(0)],$$

hence $u \in h^1$.

(iii)⇒(i). Suppose it is given that $u \in h^1$. Define

$$\mu_r(t) = \int_0^t u(re^{i\theta})d\theta,$$

then $\mu_r(0)=0$, and, for $0=t_0<t_1<\cdots<t_n=2\pi$,

$$\sum_{k=1}^{n} |\mu_r(t_k) - \mu_r(t_{k-1})| \leqslant \int_0^{2\pi} |u(re^{i\theta})|d\theta \leqslant c.$$

Hence $\{\mu_r(t)\}$ is a family of functions of uniformly bounded variation. Then by virtue of the lemma, we conclude that there exists a sequence $r_n \to 1$ such that

$$\mu_{r_n}(t) \to \mu(t) \ (n \to \infty),$$

with $\mu(t)$ being of bounded variation on $[0, 2\pi]$.

Therefore,

$$\begin{aligned} &\frac{1}{2\pi}\int_0^{2\pi} P(r,\theta\text{-}t)d\mu(t) \\ &= \lim_{n\to\infty} \frac{1}{2\pi}\int_0^{2\pi} P(r,\theta-t)d\mu_{r_n}(t) \\ &= \lim_{n\to\infty} \frac{1}{2\pi}\int_0^{2\pi} P(r,\theta-t)u(r_n e^{it})dt \\ &= \lim_{n\to\infty} u(r_n z) = u(z). \end{aligned}$$

The theorem is thus proved.

The fact that in the unit disk, every nonnegative harmonic function can be expressed as a Poisson-Stieltjes integral of nondecreasing function $\mu(t)$ is also called Herglotz representation theorem.

Also to a given $u \in h^1$, the corresponding funtion $\mu(t)$ of bounded variation is uniquely determined. Moreover, if

$$\int_0^{2\pi} P(r,\theta-t)d\mu(t) \equiv 0,$$

then

$$\int_0^{2\pi} \frac{e^{it}+z}{e^{it}-z} d\mu(t) = i\gamma \ (|z|<1, \ z = re^{i\theta}),$$

where γ is a real constant.

Using the expansion

$$\frac{e^{it}+z}{e^{it}-z}=1+2\sum_{n=1}^{\infty}e^{-int}z^n,$$

we obtain

$$\int_0^{2\pi} e^{int}d\mu(t)=0 \quad (n=0,\pm1,\pm2,\cdots).$$

As it is well-known that any characteristic function of an arbitrary interval can be approximated (in L^1 norm) by continuous periodic functions, and hence, by trignometric polynomials, it follows that $d\mu$ is a zero measure.

1.2. BOUNDARY BEHAVIORS OF POISSON-STIELTJES INTEGRALS

If $u(z)$ is the Poisson integral of a continuous and integrable function $\varphi(t)$, then to any point $t=\theta_0$, $u(z)\to\psi(\theta_0)$ as $z\to e^{i\theta_0}$. The same conclusion can be extended to Poisson-Stieltjes integrals. Moreover, when μ has continuous first derivative, then $u(z)\to\mu'(\theta_0)$ as $z\to e^{i\theta_0}$. Furthermore, if the condition on μ to be relaxed to assume that the symmetric derivative

$$D\mu(\theta_0)=\lim_{t\to0}\frac{\mu(\theta_0+t)-\mu(\theta_0-t)}{2t}$$

exists, then we can derive the following result.

Theorem 2.1.2. *Let $u(z)$ be the Poisson-Stieltjes integral with respect to μ as defined by (2.1.2) such that μ is of bounded variation, whose symmetric derivative $D\mu(\theta_0)$ exists, then the radial limit (i.e. the limit when z approaches the boundary along any radius) $\lim_{r\to1}u(re^{i\theta_0})$ exists, and*

$$\lim_{r\to1}u(re^{i\theta_0})=D\mu(\theta_0).$$

Proof. We may assume, without loss of generality, that $\theta_0=0$ and let $A=D\mu(0)$. Then

$$\begin{aligned}u(r)-A&=\frac{1}{2\pi}\int_{-\pi}^{\pi}P(r,t)[d\mu(t)-Adt]\\&=\frac{1}{2\pi}\{P(r,t)[\mu(t)-At]\}_{-\pi}^{\pi}\\&\quad-\frac{1}{2\pi}\int_{-\pi}^{\pi}[\mu(t)-At]\frac{\partial P(r,t)}{\partial t}dt.\end{aligned}$$

The first term on the right-hand side tends to zero as $r\to1$. For $0<\delta\leqslant|t|\leqslant\pi$,

$$\left|\frac{\partial P}{\partial t}\right|\leqslant\frac{2r(1-r^2)}{(1-2r\cos\delta+r^2)^2}\to0\ (r\to1).$$

Hence $u(z)-A-I_\delta\to 0$ holds for any fixed $\delta>0$, where

$$I_\delta=-\frac{1}{2\pi}\int_{-\delta}^{\delta}[\mu(t)-At]\left[\frac{\partial}{\partial t}P(r,t)\right]dt$$

$$=\frac{1}{\pi}\int_0^{\delta}\left[\frac{\mu(t)-\mu(-t)}{2t}-A\right]\cdot t\left[-\frac{\partial}{\partial t}P(r,t)\right]dt.$$

To a given $\varepsilon>0$, we may choose $\delta>0$ sufficiently small such that

$$\left|\frac{\mu(t)-\mu(-t)}{2t}-A\right|\leqslant\varepsilon\ (0<t\leqslant\delta).$$

This yields

$$|I_\delta|\leqslant\frac{\varepsilon}{2\pi}\int_{-\pi}^{\pi}t\left(-\frac{\partial P}{\partial t}\right)dt<2\varepsilon.$$

Consequently, when $r\to 1$

$$u(r)\to A.$$

Q. E. D.

According to the fact that a function of bounded variation is differentiable almost everywhere, we can derive the following two corollaries immediately.

Corollary 1. *Any function belonging to h^1 has radial limit almost everywhere.*

Corollary 2. *If $u(z)$ is the Poisson-Stieltjes integral with respect to $\psi(\in L^1)$, then $\lim_{r\to 1}u(re^{i\theta})=\psi(\theta)$, almost everywhere.*

As a further investigation, we can point out that when z approaches the boundary nontangentially from the inside of the unit disk, then, almost everywhere,

$$u(z)\to D\mu(\theta_0).$$

Also to any function that belongs to the class of bounded regular functions H^∞, its nontangential limit exists almost everywhere.

For $0<\alpha<\pi/2$, construct the sector with vertex $e^{i\theta}$, of angle 2α, symmetric with respect to the ray from the origin through $e^{i\theta}$. Draw the two segments from the origin perpendicular to the boundaries of this sector, and let $S_\alpha(\theta)$ denote the so formed region.

Theorem 2.1.3. *If $f\in H^\infty$, then the radial limits*

$$\lim_{r\to 1}f(re^{i\theta})$$

exist almost everywhere, moreover, if at θ_0

$$\lim_{r\to 1} f(re^{i\theta_0})$$

exists, then as $z\to e^{i\theta_0}$ in an arbitrary sector $S_a(\theta_0)$ $(0<a<\pi/2)$, $f(z)$ assumes the same limit.

Proof. Since $h^\infty \subset h^1$, it follows immediately from Corollary 1 of Theorem 2. 1. 2 that the radial limits exist almost everywhere.

In order to consider the "sectorial" limits, we shall treat functions $f(z)$, which are regular in the circle $|z-1|<1$ and it has a limit L as z approaching the origin along the positive real axis.

Let

$$f_n(z)=f(\frac{z}{n}) \quad (n=1,2\cdots),$$

then $\{f_n(z)\}$ is uniformly bounded, hence it forms a normal family. It follows that there exists a subsequence of functions $\{f_{n_k}(z)\}$, which converges uniformly to a regular function $F(z)$ in any closed subdomain of the above mentioned circle.

Since $f_n(z)\to L$ for any real z satisfying $0<z<2$, it will yield $F(z)\equiv L$. Furthermore, for z in the region:

$$|Arg\ z|\leqslant a<\frac{\pi}{2}, \quad \frac{\cos a}{2}\leqslant|z|\leqslant\cos a \tag{2. 1. 3}$$

(the ray $Arg\ z=a$, that lies inside of the circle: $|z-1|<1$ has a length of $2\cos a$), we have $f_n(z)\to L$ uniformly as $n\to\infty$. Hence

$$f(z)\to L \quad (z\to 0,\ z\in\{z: |Arg\ z|\leqslant a\}),$$

the theorem is thus proved.

If $f(z)$ tends to L as z approaching $e^{i\theta_0}$ in any given sector $S_a(\theta_0)(a<\pi/2)$, then we call $f(z)$ having a non-tangential limit L (at point $e^{i\theta_0}$).

We can conclude from the above argument that to any function $f(z)\in H^\infty$, its non-tangential limits exist almost everywhere (on the boundary of the unit circle).

1. 3. SUBHARMONIC FUNCTIONS

Let D denote a bounded domain, ∂D its boundary, and $\overline{D}$ the closure of D. Let $g(z)$ be a real-valued continuous function defined in D (the value ∞ shall also be allowed, but the function which is identically equal to ∞ shall be excluded). Suppose that for any arbitrary subdomain B with $\overline{B} \subset D$, and a function $U(z)$ which is harmonic in B and continuous on $\overline{B}$, if

$$g(z) \leqslant U(z) \quad (z \in \partial B),$$

implying $g(z) \leqslant U(z)$ in B, then any function assumes the same properties as $g(z)$ described above is called a subharmonic function in D.

Theorem 2.1.4. *Let $g(z)$ be a real-valued continuous function defined in a domain D. A necessary and sufficient condition for $g(z)$ to be subharmonic in D is that to every $z_0 \in D$, there is a $\rho_0 > 0$ (depending on z_0) with $|z-z_0| < \rho_0 \subset D$ (if $z_0 \neq \infty$) such that the following inequality holds:*

$$g(z_0) \leqslant \frac{1}{2\pi} \int_0^{2\pi} g(z_0+\rho e^{i\theta})d\theta, \text{ for all } \rho < \rho_0,$$

while for $z_0 = \infty$

$$g(z_0) \leqslant \frac{1}{2\pi} \int_0^{2\pi} g(z_0+\rho e^{i\theta})d\theta, \text{ for all } \rho > \rho_0. \tag{2.1.4}$$

Proof. We shall prove the case that $z_0 \neq \infty$. The necessity is easy to show. Assume that the disk $|z-z_0| < \rho$ contained in D and $U(z)$ is a harmonic function defined in there satisfying

$$U(z) = g(z) \qquad (|z-z_0| = \rho),$$

then

$$g(z) \leqslant U(z_0) = \frac{1}{2\pi} \int_0^{2\pi} U(z_0+\rho e^{i\theta})d\theta$$

$$= \frac{1}{2\pi} \int_0^{2\pi} g(z_0+\rho e^{i\theta})d\theta.$$

Now let us prove its sufficiency. Let B be an arbitrary domain in D with $\overline{B} \subset D$. Assume that there exists a harmonic function $U(z)$ satisfying $g(z) \leqslant U(z)$ $(z \in \partial B)$. Suppose that there exists a certain point z in B satisfying $g(z) > U(z)$. Set

$$h(z) = g(z) - U(z),$$

and assume that on $\overline{B}$, $h(z)$ attains its maximal value $m(>0)$ on a subset $E \subset \overline{B}$. Since $h(z) \leqslant 0$ for $z_0 \in \partial B$, it follows that $E \subset B$. Moreover, because E is a closed set, there exists some z_0 in E such that none of its disklike neighborhood can be contained in B completely. Hence there exists a sequence of numbers $\{\rho_n\}$ with $\rho_n > 0$ such that each of the disk: $|z-z_0| < \rho_n$ contained in B, but the circumference $|z-z_0| = \rho_n$ fails to lie completely in E, i.e., on part of the circumference $h(z)<m$. It follows that

$$\frac{1}{2\pi}\int_0^{2\pi} g(z_0+\rho_n e^{i\theta})d\theta - U(z_0)$$
$$=\frac{1}{2\pi}\int_0^{2\pi} h(z_0+\rho_n e^{i\theta})d\theta$$
$$< m = h(z_0) = g(z_0)-U(z_0).$$

This contradicts with (2.1.4). Theorem 2.1.4 is thus proved.

Example 1. If $f(z)$ is regular in a domain D and $p>0$, then $g(z)=|f(z)|^p$ is a subharmonic function in D.

In fact, if $f(z_0)=0$, then the conclusion holds naturally. If $f(z_0)\neq 0$, then $f(z)$ is regular in a neighborhood of z_0. Hence

$$[f(z_0)]^p=\frac{1}{2\pi}\int_0^{2\pi}[f(z_0+\rho e^{i\theta})]^p d\theta \qquad \text{(for sufficiently small } \rho\text{)}.$$

Consequently

$$|f(z_0)|^p \leqslant \frac{1}{2\pi}\int_0^{2\pi}|f(z+\rho e^{i\theta})|^p d\theta \qquad (\rho \text{ sufficiently small}).$$

It follows from Theorem 2.1.4 that $|f(z)|^p$ is subharmonic in D.

Example 2. If $u(z)$ is harmonic in D and $p\geqslant 1$, then $|u(z)|^p$ is subharmonic in D.

It is obvious when $p=1$. *If* $p>1$, then, by Hölder's inequality, we have

$$|u(z_0)|\leqslant\frac{1}{2\pi}\left[\int_0^{2\pi}|u(z_0+\rho e^{i\theta})|^p d\theta\right]^{\frac{1}{p}}(2\pi)^{\frac{1}{q}},$$

where $\frac{1}{p}+\frac{1}{q}=1$. Inequality (2.1.4) is thus satisfied.

Example 3. If $f(z)$ is a regular function defined in a domain D, then $g(z)=\log^+|f(z)|$ is a subharmonic function in D.

In fact, when $|f(z_0)|<1$, condition (2.1.4) holds obviously. When $|f(z_0)|>1$, then the inequality holds in a neighborhood of z_0: $|z-z_0|<\rho$. Furthermore,

$$\log^+|f(z)| = \log|f(z)|,$$

and note that $\log|f(z)|$ is harmonic there.

1.4. THE CONVEXITY THEOREM OF HARDY

In the following we shall consider functions that belong to H^2. If

$$f(z)=\sum_{n=0}^{\infty}a_n z^n \quad (|z|<1),$$

is regular in $|z|<1$, then, according to Parseval formula from mathematical analysis,

$$M_2^2(r,f)=\sum_{n=0}^{\infty}|a_n|^2 r^{2n}.$$

It follows from this that $M\ (r,f)$ is an increasing function of r. Moreover, a necessary and sufficient condition for a function $f(z)$ to be in H^2 is that $\sum_{n=0}^{\infty}|a_n|^2<\infty$. Also from the principle of maximal modulus one can derive that $M_\infty(r,f)$ is an increasing function of r. In general the following results hold:

Theorem 2.1.5. (Hardy's convexity theorem). *Let $f(z)$ be regular in $|z|<1$ and $0<p\leqslant\infty$, then*

(i) *$M_p(r,f)$ is a non-decreasing function of r;*

(ii) *$\log M_p(r,f)$ is a convex function of $\log r$.*

In order to prove the theorem, we prove first the following result.

Theorem 2.1.6. *Let $g(z)$ be a subharmonic function defined in $|z|<1$. Define*

$$m(r)=\frac{1}{2\pi}\int_0^{2\pi}g(re^{i\theta})d\theta \quad (0\leqslant r<1),$$

then $m(r)$ is non-decreasing and is a convex function of $\log r$.

Proof. Choose $0\leqslant r_1<r_2<1$. Let $U(z)$ be a function that is in $|z|<r_2$ and continuous on $|z|\leqslant r_2$ such that

$$U(z)=g(z) \quad (|z|=r_2).$$

Now since $g(z)$ is subharmonic, it follows that

$$g(z)\leqslant U(z) \quad (|z|\leqslant r_2).$$

Hence

$$m(r_1)\leqslant\frac{1}{2\pi}\int_0^{2\pi}U(r_1e^{i\theta})d\theta=U(0)$$

$$=\frac{1}{2\pi}\int_0^{2\pi}U(r_2e^{i\theta})d\theta=m(r_2).$$

Now we proceed to prove the convexity part of the theorem. Let $0<r_1<r_2<1$ and $U(z)$ be a harmonic function defined in $r_1<|z|<r_2$. Then

$$m(r)\leqslant\frac{1}{2\pi}\int_0^{2\pi}U(re^{i\theta})\,d\theta\ (r_1\leqslant r\leqslant r_2), \tag{2.1.5}$$

and the equal sign holds at the two end points.

On the other hand, from Green's formula,

$$\frac{d}{dr}\left\{\int_0^{2\pi}U(re^{i\theta})d\theta\right\}$$

$$=\frac{1}{r}\int_0^{2\pi}\frac{\partial U}{\partial n}ds=\frac{c}{r},$$

where $\frac{\partial}{\partial n}$ denotes the normal derivative, $ds=r\,d\theta$ the element of arc length, c a constant. It follows that the right side of (2.1.5) assumes the form:

$$a\log r+b.$$

We say that $m(r)$ is a convex function of $\log r$, meaning that if

$$\log r=a\log r_1+(1-a)\log r_2$$
$$(0<r_1<r_2<1;\ 0<a<1),$$

then

$$m(r)\leqslant a\,m(r_1)+(1-a)m(r_2).$$

Now from (2.1.5) and the form of its right-hand side, we have

$$\begin{aligned}m(r)&\leqslant a\log r+b\\ &=a\log(r_1^a r_2^{1-a})+b\\ &=a(a\log r_1+b)+(1-a)(a\log r_2+b)\\ &=a\Big[\frac{1}{2\pi}\int_0^{2\pi}U(r_1e^{i\theta})d\theta\Big]\\ &\quad+(1-a)\Big[\frac{1}{2\pi}\int_0^{2\pi}U(r_2e^{i\theta})d\theta\Big]\\ &=a\,m(r_1)+(1-a)m(r_2).\end{aligned}$$

This also completes the proof of Theorem 2. 1. 6.

We now proceed the proof of Hardy's convexity theorem as follows.

(i). It has been pointed out in Section 1. 3 that if $f(z)$ is regular in D, then $g(z)=|f(z)|^p$ is subharmonic in D.

Similarly, if $u(z)$ is harmonic, then $|u(z)|^p(p\geqslant1)$ is subharmonic. Thus by applying Theorem 2. 1. 6 we can derive that $M_p(r,f)$ is a nondecreasing function of r.

(ii). It has been pointed out at the conclusion of Section 1. 3 that, for $p>0$ and $f(z)$ regular in $0<|z|<1$, the function $g(z)=|z|^{\lambda}|f(z)|^{p}$ is subharmonic in $0<|z|<1$. It follows from Theorem 2. 1. 6 that $r^{\lambda}M_p^p(r,f)$ is a convex function of $\log r$.

Now given $0<r_1<r_2<1$, choose $\lambda<0$ such that

$$r_1^{\lambda} M_p^p(r_1,f)=r_2^{\lambda}M_p^p(r_2,f)=k.$$

Set

$$r=r_1^a r_2^{1-a} \quad (0<a<1),$$

then

$$r_{\lambda}M_p^p(r,f)\leqslant k=k^a k^{1-a}=\{r_1^{\lambda}M_p^p(r_1,f)\}^a\{r_2^{\lambda}M_p^p(r_2,f)\}^{1-a}$$
$$=r^{\lambda}\{M_p^p(r_1,f)\}^a\{M_p^p(r_2,f)\}^{1-a}.$$

The assertion that $\log M(r,f)$ is a convex function of $\log r$ follows immediately by taking the logarithm on both sides of the above inequality.

1. 5. SUBORDINATION

Let $f(z)$ and $F(z)$ be two univalent functions regular in $|z|<1$ with $f(0)=0$, $F(0)=0$. If the range (image) set of F contains the range set of $f(z)$, then $w(z)=F^{-1}(f(z))$ is regular with $w(0)=0$ and $|w(z)|<1$ in $|z|<1$. Further, by Schwarz lemma, $|w(z)|\leqslant|z|$, It follows particularly that in every disk $|z|\leqslant r<1$, the range set of $f(z)=F(w(z))$ is contained in that of $F(z)$.

More generally (without the requirement of the univalency), if there exists a function $w(z)$ regular in $|z|<1$ satisfying $|w(z)|\leqslant|z|$ such that $f(z)=F(w(z))$, where f and F both are regular in $|z|<1$, then we shall call f subordinate to F and will denote such a relation by $f\prec F$.

Theorem 2. 1. 7. (Littlewood). *Let $f(z)$ and $F(z)$ be regular in $|z|<1$. If $f\prec F$, then*

$$M_p(r,f)\leqslant M_p(r,F) \quad (0<p\leqslant\infty).$$

Proof. Let $G(z)$ be a function subharmonic in $|z|<1$. Consider $g(z)=G(w(z))$, where $w(z)$ is regular with $|w(z)|<|z|$ in $|z|<1$. Then

$$\int_0^{2\pi} g(re^{i\theta})\,d\theta\leqslant\int_0^{2\pi} G(re^{i\theta})d\theta. \tag{2. 1. 6}$$

In fact, let $U(z)$ be harmonic in $|z|<r$ with $G(z)=U(z)$ on the unit circle: $|z|=1$, then

$$G(z)\leqslant U(z) \qquad (|z|\leqslant r).$$

Hence

$$g(z) \leqslant u(z) = U(w(z)) \qquad (|z|=r).$$

Therefore

$$\frac{1}{2\pi}\int_0^{2\pi} g(re^{i\theta})d\theta \leqslant \frac{1}{2\pi}\int_0^{2\pi} u(re^{i\theta})d\theta = u(0) = U(0)$$

$$= \frac{1}{2\pi}\int_0^{2\pi} U(re^{i\theta})\, d\theta = \frac{1}{2\pi}\int_0^{2\pi} G(re^{i\theta})d\theta.$$

The assertion of the theorem will be obtained by noting the facts that $|f(z)|^p$, $|F(z)|^p$ are subharmonic, and then followed by applying inequality (2.1.6) letting $g(re^{i\theta}) = f(re^{i\theta})$, $G(re^{i\theta}) = F(re^{i\theta})$.

2. The Basic Structure of H^p

2.1. BOUNDARY VALUES

Definition. We shall use N to denote the class of functions $f(z)$ that are regular in the unit disk and satisfy

$$\sup_{r<1}\int_0^{2\pi} \log^+ |f(re^{i\theta})|\, d\theta < \infty.$$

Since the integration

$$\exp\left\{\frac{1}{2\pi}\int_0^{2\pi} \log(1+|f(re^{i\theta})|)d\theta\right\}$$

is no greater than the integration $\{\frac{1}{2\pi}\int_0^{2\pi}[1+|f(re^{i\theta})|]^p d\theta\}^{\frac{1}{p}}$ (where p is an arbitrary positive number), it follows that

$$H^p \subset N \quad (p>0).$$

Theorem 2.2.1. (F. and R. Nevanlinna). *A necessary and sufficient condition for a function $f(z)$ regular in the unit disk belongs to the class N is that $f(z)$ can be expressed as*

$$f(z) = \frac{\varphi(z)}{\psi(z)},$$

where φ and ψ are two functions that are regular and bounded in the unit disk.

Proof. We first prove the sufficiency. We may assume without any loss of the generality that

$$|\varphi(z)|\leqslant 1,\ |\psi(z)|\leqslant 1,\ \psi(0)\neq 0.$$

Then

$$\int_0^{2\pi} \log^+|f(re^{i\theta})|\,d\theta \leqslant -\int_0^{2\pi} \log|\psi(re^{i\theta})|\,d\theta.$$

Hence, by virtue of Jensen's formula,

$$\frac{1}{2\pi}\int_0^{2\pi} \log|\psi(re^{i\theta})|\,d\theta = \log|\psi(0)| + \sum_{|z_n|<r} \log\frac{r}{|z_n|},$$

where z_n are the zeros of ψ. It shows clearly that $\int_0^{2\pi} \log|\psi|\,d\theta$ is an increasing function of r, and, hence, $f\in N$.

Proof of the necessity. Assume $f(z)\not\equiv 0$ and $f\in N$. *If* $z=0$ is a zero of f with multiplicity $m(\geqslant 0)$, then $z^{-m}f(z)\to a\ (\neq 0)\ (z\to 0)$. Let $z_n\,(n=1,2,3,\cdots)$ denote the rest zeros of f with

$$0<|z_1|\leqslant|z_2|\leqslant\cdots<1.$$

We know that if $f(z)\neq 0$ on the circle $|z|=\rho\ (<1)$, then

$$F(z)=\log\{f(z)\frac{\rho^m}{z^m}\prod_{|z_n|<\rho}(\frac{\rho^2-\bar{z}_n z}{\rho(z-z_n)})\}$$

is a function that is regular in $|z|\leqslant\rho$ and satisfies

$$Re\{F(z)\}=\log|f(z)|\quad(|z|=\rho).$$

Hence, by Poisson's formula,

$$F(z)=\frac{1}{2\pi}\int_0^{2\pi}\log|f(\rho e^{it})|\frac{\rho e^{it}+z}{\rho e^{it}-z}dt+ic.$$

Thus we can express $f(z)$ as $\dfrac{\varphi_\rho(z)}{\psi_\rho(z)}$, where

$$\varphi_\rho(z)=\frac{z^m}{\rho^m}\prod_{|z_n|<\rho}\frac{\rho(z-z_n)}{\rho^2-\bar{z}_n z}\cdot exp\{-\frac{1}{2\pi}\int_0^{2\pi}\log^-|f(\rho e^{i\theta})|\frac{\rho e^{it}+z}{\rho e^{it}-z}dt+ic\},$$

$$\psi_\rho(z)=exp\{-\frac{1}{2\pi}\log^+|f(\rho e^{it})|\frac{\rho e^{it}+z}{\rho e^{it}-z}dt\},$$

here $\log^- x=\max(-\log x,0)$.

Now we choose a sequence of real numbers: $\{\rho_k\}$ with ρ_k increasing monotonically to 1, such that $f(z)\neq 0$ on $|z|=\rho_k$.

Let

$$\varphi_k(z)=\varphi_{\rho_k}(\rho_k z);\ \psi_k(z)=\psi_{\rho_k}(\rho_k z),$$

then clearly

$$f(\rho_k z)=\frac{\varphi_k(z)}{\psi_k(z)}\qquad(|z|<1).$$

Note that φ_k, ψ_k are regular in the unit disk D and satisfy $|\varphi_k(z)| \leqslant 1$, $|\psi_k(z)| < 1$ in D. It follows that $\{\varphi_k\}$, $\{\psi_k\}$ are normal families, and, hence, there exists a subsequence $\{k_j\}$ of the natural numbers such that when $j \to \infty$

$$\varphi k_j(z) \overset{uniformly}{\to} \varphi(z) \quad (|z| \leqslant R < 1),$$

$$\psi k_j(z) \overset{uniformly}{\to} \psi(z) \quad (|z| \leqslant R < 1).$$

It follows that both $\varphi(z)$ and $\psi(z)$ are regular in the unit disk D and satisfy $|\varphi(z)| \leqslant 1$, $|\psi(z)| \leqslant 1$ in D. Hence

$$f(z) = \frac{\varphi(z)}{\psi(z)}.$$

The theorem is thus proved.

Theorem 2. 2. 2. *Let $f \in N$. Then non-tangential limit $f(e^{i\theta})$ exists almost everywhere, and $\log|f(e^{i\theta})|$ is integrable (except $f(z) \equiv 0$). Furthermore, if $f \in H^p$, then $f(e^{i\theta}) \in L^p$.*

Proof. Assume that $f(z) \not\equiv 0$ and that $f \in N$. Then

$$f(z) = \frac{\varphi(z)}{\psi(z)} \quad (|\varphi(z)| \leqslant 1, |\psi(z)| \leqslant 1).$$

where $\varphi(z)$ and $\psi(z)$ are two bounded and regular functions in the unit disk, having non-tangential limits $\varphi(e^{i\theta})$, $\psi(e^{i\theta})$ almost everywhere. Applying Theorem 2. 1. 3 to Fatou′ s lemma, we obtain

$$\int_0^{2\pi} |\log|\varphi(e^{i\theta})|\, d\theta$$

$$\leqslant \liminf_{r \to 1} \{-\int_0^{2\pi} \log|\varphi(re^{i\theta})|\, d\theta\}.$$

It follows from Jensen′ s theorem that $\int_0^{2\pi} \log|\varphi(re^{i\theta})|\, d\theta$ is an increasing function of r, and, hence $\log|\varphi(e^{i\theta})| \in L^1$; a similar conclusion will be held for ψ.

Particularly, $\psi(e^{i\theta})$ cann′ t be zero identically on a set (of θ) of positive measure. Hence radial limit $f(re^{i\theta})$ exists almost everywhere, and, furthermore, $\log|f(e^{i\theta})| \in L^1$. Moreover, if $f \in H^p$, then, by applying Fatou′ s lemma, one can readily verify that $f(e^{i\theta}) \in L^p$.

The theorem also shows that if $f \in N$ and $f(e^{i\theta}) = 0$ on a set of positive measure, then $f(z) \equiv 0$. It also reveals that any function in the class N is uniquely determined by its boundary values on a set of arguments of positive measure.

From the expression $f=\frac{\varphi}{\psi}$, it shows clearly that if $f\in N$, then $\int_0^{2\pi} \log^- |f(re^{i\theta})|d\theta$ is bounded. Thus, a necessary and sufficient condition for a function $f\in N$ is that $\int_0^{2\pi} |\log|f(re^{i\theta})||d\theta<\infty$, where r satisfies $0<r<1$.

2.2. ZEROS

Theorem 2.2.3. *Let f be regular in $|z|<1$ with $f\not\equiv 0$. Let $z_1, z_2, \cdots$ be the zeros of f with $0<|z_1|\leqslant|z_2|\leqslant\cdots$. Then a necessary and sufficient condition for $\int_0^{2\pi} \log|f(re^{i\theta})|d\theta$ to be bounded is that*

$$\sum_{n=1}^{\infty}(1-|z_n|)<\infty.$$

Proof. Assume that $z=0$ is a zero of f with multiplicity m. Then

$$f(z)=dz^m+\cdots \quad (d\neq 0).$$

The rest zeros of f are denoted by $a_n(n=1,2,\cdots)$ with $0<|a_1|\leqslant|a_2|\leqslant\cdots$. According to Jensen's formula,

$$\frac{1}{2\pi}\int_0^{2\pi} \log|f(re^{i\theta})|d\theta=\sum_{|a_n|<r} \log\frac{r}{|a_n|}+\log(|d|r^m),$$

which is an increasing function of r; if it is bounded by a constant c (finite), then, for any fixed integer k,

$$\begin{aligned}\sum_{n=1}^{k} \log\frac{r}{|a_n|}&\leqslant\sum_{|a_n|<r} \log\frac{r}{|a_n|} \quad (r>|a_k|)\\ &\leqslant c-\log(|d|r^m).\end{aligned}$$

Letting $r\to 1$, we obtain

$$0<|d|e^{-c}\leqslant\prod_{n=1}^{k}|a_n|,$$

it follows from this that $\sum(1-|a_n|)<\infty$.

Conversely, by applying Jensen's theorem, we have

$$\Big(\prod_{|a_n|<r}|a_n|\Big)\exp\{\frac{1}{2\pi}\int_0^{2\pi} \log|f(re^{i\theta})|d\theta\}<|d|,$$

and the assertion of the theorem follows.

Corollary. *The zero sets of functions in N must satisfy*

$$\sum (1-|z_n|)<\infty.$$

Theorem 2. 2. 4. *Let $a_1, a_2, \cdots$, be a sequence of complex numbers such that $0<|a_1|\leqslant |a_2|\leqslant\cdots<1$ and $\sum_{n=1}^{\infty}(1-|a_n|)<\infty$, then the infinite product*

$$B(z)=\prod_{n=1}^{\infty}\frac{|a_n|}{a_n}\quad\frac{a_n-z}{1-\bar{a}_n z}$$

converges uniformly in every disk: $|z|\leqslant R<1$ with $\{a_n\}$ being the sole zeros of $B(z)$ in the unit disk. Moreover, $|B(z)|<1$ $(|z|<1)$ and $|B(e^{i\theta})|=1$ almost everywhere on the unit circle.

Proof. For $|z|\leqslant R$, we have

$$\left|1-\frac{|a_n|}{a_n}\quad\frac{a_n-z}{1-\bar{a}_n z}\right|=\left|\frac{(a_n+|a_n|z)(1-|a_n|)}{a_n(1-\bar{a}_n z)}\right|$$
$$\leqslant\frac{2(1-|a_n|)}{1-R}.$$

Since $\sum(1-|a_n|)<\infty$, it follows that the infinite product $B(z)$ converges uniformly in $|z|\leqslant R<1$ and is regular in $|z|<1$.

When $|z|<1$, the modulus of any finite product of $B(z)$ is less than 1. From this we conclude easily that $|B(z)|<1$ and the radial limits of $B(z)$ exist almost everywhere with $|B(e^{i\theta})|\leqslant 1$.

We now show that actually

$$|B(e^{i\theta})|=1 \text{ almost everywhere.}$$

To any $f\in H$, we have, from Lebesgue' s bounded convergence theorem and the monotonicity of $M_1(r,f)$,

$$\int_0^{2\pi}|f(re^{i\theta})|d\theta\leqslant\int_0^{2\pi}|f(e^{i\theta})|d\theta.$$

We apply the above result to $f=\frac{B}{B_n}$, where

$$B_n(z)=\prod_{k=1}^{n}\frac{|a_k|}{a_k}\,\frac{a_k-z}{1-\bar{a}_k z}.$$

Since $|B_n(e^{i\theta})|\equiv 1$, it follows that

$$\int_0^{2\pi}\left|\frac{B(re^{i\theta})}{B_n(re^{i\theta})}\right|d\theta\leqslant\int_0^{2\pi}|B(e^{it})|dt.$$

But

$$B_n(z) \xrightarrow{\text{uniformly}} B(z) \qquad (|z|=r).$$

Hence

$$2\pi \leqslant \int_0^{2\pi} |B(e^{it})|dt.$$

From this and the fact that $|B(e^{i\theta})| \leqslant 1)$ (almost everywhere), we conclude that it must be $|B(e^{i\theta})|=1$ almost everywhere, and the proof is complete.

In general, the expression:

$$B(z)=z^m \prod_{n=1}^{\infty} \frac{|a_n|}{a_n} \frac{a_n-z}{1-\bar{a}_n z},$$

where m is a non-negative integer with

$$\sum (1-|a_n|)<\infty,$$

is called a Blaschke product.

2. 3. THE MEANING OF CONVERGING TO THE BOUNDARY VALUES

Theorem 2. 2. 5. (F. Riesz). *Any function $f(z)(\not\equiv 0)$ that belongs to $H^p(p>0)$ can be decomposed as*

$$f(z)=B(z)g(z),$$

where $B(z)$ is the Blaschke product, $g(z)\in H^p$ and has no zeros in $|z|<1$.

Similarly, to every $f\in N$, it can be decomposed as

$$f=Bg,$$

where g belongs to N and does not vanish in $|z|<1$.

Proof. We may assume that f has infinitely many zeros (otherwise, the conclusion is obvious).

Let

$$B_n(z)=z^m \prod_{k=1}^{n} \frac{|a_k|}{a_k} \frac{a_k-z}{1-\bar{a}_k z}, \quad g_n(z)=\frac{f(z)}{B_n(z)}.$$

Then for a fixed n and $\varepsilon>0$ with $|z|$ being sufficiently close to 1, we have $|B_n(z)|>1-\varepsilon$. Hence

$$\int_0^{2\pi} |g_n(re^{i\theta})|^p d\theta \leqslant (1-\varepsilon)^{-p} \int_0^{2\pi} |f(re^{i\theta})|^p d\theta$$

$$\leqslant \frac{M}{(1-\varepsilon)^p}.$$

The value of the integration is thus bounded by a quantity which is independent of r and decreases monotonically with r. From the above, by letting $\varepsilon \to 0$, we obtain

$$\int_0^{2\pi} |g_n(re^{i\theta})|^p d\theta \leqslant M \quad (n, r(<1) \text{ arbitrary}). \tag{2.2.1}$$

By virtue of Theorem 2.2.4 we have by letting $n \to \infty$,

$$g_n(z) \overset{uniformly}{\to} g(z) = \frac{f(z)}{B(z)} \quad (|z| = R < 1).$$

This shows that $g \in H^p$ and has no zeros in $|z| < 1$, and the theorem is thus proved.

The proof for $f \in N$ is similar.

Theorem 2.2.6. *If $f \in H^p (0 < p < \infty)$, then*

$$\lim_{r \to 1} \int_0^{2\pi} |f(re^{i\theta})|^p d\theta = \int_0^{2\pi} |f(e^{i\theta})|^p d\theta. \tag{2.2.2}$$

Furthermore

$$\lim_{r \to 1} \int_0^{2\pi} |f(re^{i\theta}) - f(e^{i\theta})|^p d\theta = 0. \tag{2.2.3}$$

Proof. Firstly we prove the assertion (2.2.3) when $p = 2$.

Since

$$f(z) = \sum_{n=0}^{\infty} a_n z^n \in H^2,$$

then

$$\sum_{n=0}^{\infty} |a_n|^2 < \infty.$$

It follows from Fatou's lemma that

$$\int_0^{2\pi} |f(re^{i\theta}) - f(e^{i\theta})|^2 d\theta \leqslant \liminf_{\rho \to 1} \int_0^{2\pi} |f(re^{i\theta}) - f(\rho e^{i\theta})|^2 d\theta$$

$$= 2\pi \sum_{n=1}^{\infty} |a_n|^2 (1 - r^n)^2.$$

Hence when $r \to 1$, $\int_0^{2\pi} |f(re^{i\theta}) - f(e^{i\theta})|^2 d\theta \to 0$; this shows that (2.2.3) is true and so does (2.2.2) for $p = 2$.

Now consider $f \in H^p (0 < p < \infty)$. According to Theorem 2.2.5 we decompose f into $f(z) = B(z) g(z)$. Since $[g(z)]^{\frac{p}{2}} \in H^2$, hence

$$\int_0^{2\pi}|f(re^{i\theta})|^p d\theta \leqslant \int_0^{2\pi}|g(re^{i\theta})|^p d\theta$$

$$\to \int_0^{2\pi}|g(e^{i\theta})|^p d\theta = \int_0^{2\pi}|f(e^{i\theta})|^p d\theta.$$

Applying Fatou' s lemma again, we obtain assertion (2. 2. 3).

Next by using the lemma below, (2. 2. 3) can be derived from (2. 2. 2) immediately.

Lemma 1. *Let Ω be a measurable subset of the real axis and $\varphi_n \in L^p(\Omega)$, $0<p<\infty$; $n=1,2,\cdots$. Suppose that $\varphi_n(x)\to\varphi(x)$ (for $x\in\Omega$ near every-where, $n\to\infty$) and*

$$\int_\Omega |\varphi_n(x)|^p dx \to \int_\Omega |\varphi(x)|^p dx < \infty (n\to\infty).$$

Then

$$\int_\Omega |\varphi_n(x)-\varphi(x)|^p dx \to 0 \ (n\to\infty).$$

Proof. To a measurable set $\Omega'\subset\Omega$, we define

$$J_n(\Omega') = \int_{\Omega'} |\varphi_n|^p dx,$$

$$J(\Omega') = \int_{\Omega'} |\varphi|^p dx,$$

$$\widetilde{\Omega}' = \Omega - \Omega'.$$

Then

$$J(\Omega') \leqslant \liminf_{n\to\infty} J_n(\Omega') \leqslant \limsup_{n\to\infty} J_n(\Omega')$$
$$\leqslant \lim_{n\to\infty} J_n(\Omega) - \liminf_{n\to\infty} J_n(\widetilde{\Omega}')$$
$$\leqslant J(\Omega) - J(\widetilde{\Omega}') = J(\Omega').$$

Hence, to any measurable set $\Omega'\subset\Omega$,

$$J_n(\Omega') \to J(\Omega') \ (n\to\infty).$$

To a given $\varepsilon>0$, we can choose a set $\Omega_0\subset\Omega$ of finite measure such that

$$J(\Omega_0)<\varepsilon.$$

Now choose $\delta>0$ such that to every measurable set $Q\subset\Omega_0$ it will satisfy $J(Q)<\varepsilon$, whenever $m(Q)<\varepsilon$. By Egorov' s theorem, there exists a set $Q\subset\Omega_0$ (with $m(Q)<\delta$) such that

$$\varphi_n(x)\to\varphi(x) \text{ uniformly on } \Omega' = \Omega_0 - Q \text{ as } n\to\infty.$$

Since

$$J_n(\Omega_0)\to J(\Omega_0) \qquad (n\to\infty),\ J_n(Q)\to J(Q) \qquad (n\to\infty),$$

it follows that, for sufficiently large n,

$$\int_\Omega |\varphi_n-\varphi|^p dx=\int_{\Omega_0} |\varphi_n-\varphi|^p dx$$

$$+\int_{\Omega'} |\varphi_n-\varphi|^p dx+\int_Q |\varphi_n-\varphi|^p dx\leqslant 2^p\{J_n(\Omega_0)$$

$$+J(\Omega_0)+J_n(Q)+J(Q)\}+\int_{\Omega'} |\varphi_n-\varphi|^p dx<(2^p6+1)\varepsilon.$$

Lemma 1 is thus proved, and, hence, Theorem 2. 2. 6 follows.

Corollary. *If* $f\in H^p(p>0)$, *then*

$$\lim_{r\to 1}\int_0^{2\pi} |\log^+|f(re^{i\theta})|-\log^+|f(e^{i\theta})||d\theta=0.$$

The above corollary can be derived directly from Theorem 2. 2. 6 and Lemma 2 below.

Lemma 2. *If* $a\geqslant 0$, $b\geqslant 0, 0<p\leqslant 1$, *then*

$$|\log^+a-\log^+b|\leqslant\frac{1}{p}|a-b|^p.$$

Proof. It suffices to prove the lemma under the condition: $1\leqslant b<a$.

By differential calculus, we can verify easily the following inequality:

$$\log x\leqslant\frac{1}{p}(x-1)^p \qquad (x\geqslant 1).$$

Put $x=a/b$, the assertion follows.

To $1\leqslant p<\infty$, let

$$\|f\|_p=\{\frac{1}{2\pi}\int_0^{2\pi} |f(e^{i\theta})|^p d\theta\}^{\frac{1}{p}}=\lim_{r\to 1} M_p(r,f);$$

and to $p=\infty$, let

$$\|f\|_\infty=\sup_{|z|<1}|f(z)|=\operatorname{ess}\sup_{0\leqslant\theta<2\pi}|f(e^{i\theta})|,$$

then it is easily seen that H^p forms a normed linear space.

2. 4. CANONICAL FACTORIZATION

Theorem 2. 2. 7. *If* $f\in H^p$, $p<0$, *then*

$$\log|f(re^{i\theta})|\leqslant\frac{1}{2\pi}\int_0^{2\pi}P(r,\theta-t)\log|f(e^{it})|dt.$$

Proof. By Theorem 2. 2. 5 we may assume $f(z)\neq 0$ inside the unit disk $|z|<1$. Thus $\log|f(z)|$ is harmonic in $|z|<1$, and hence,

$$\log|f(\rho re^{i\theta})|=\frac{1}{2\pi}\int_0^{2\pi}P(r,\theta-t)\log|f(\rho e^{it})|dt \qquad (r<\rho<1),$$

it follows from the corollary of Theorem 2. 2. 6 that

$$\lim_{\rho\to 1}\int_0^{2\pi}P(r,\theta-t)\log^+|f(\rho e^{it})|dt=\int_0^{2\pi}P(r,\theta-t)\log^+|f(e^{it})|dt$$

On the other hand, by Fatou' s lemma

$$\lim_{\rho\to 1}\int_0^{2\pi}P(r,\theta-t)\log^-|f(\rho e^{it}|dt\geqslant\int_0^{2\pi}P(r,\theta-t)\log^-|f(e^{it})|dt.$$

Now by combining the above two results the assertion follows.

We now discuss the factorization problem, assuming that $f(z)\not\equiv 0$, $f\in H^p(p>0)$. According to Theorem 2. 2. 2, it follows that $f(e^{i\theta})\in L^p$ and $\log|f(e^{i\theta})|\in L^1$.

Consider the entire function

$$f(z)=\exp\left\{\frac{1}{2\pi}\int_0^{2\pi}\frac{e^{it}+z}{e^{it}-z}\log|f(e^{it})|dt\right\}. \qquad (2.\ 2.\ 4)$$

According to Theorem 2. 2. 5, we have $f(z)=B(z)g(z)$; $g(z)\neq 0$. Furthermore by Theorem 2. 2. 7, $|g(e^{i\theta})|=|f(e^{i\theta})|$ (almost everywhere) and $|g(z)|\leqslant|F(z)|$ $(|z|<1)$. By virtue of Corollary 2 of Theorem 2. 1. 2, we have

$$|g(e^{i\theta})|=|F(e^{i\theta})| \text{ (almost everywhere)}$$

Set $e^{i\gamma}=\dfrac{g(0)}{|g(0)|}$ and $S(z)=\dfrac{e^{i\gamma}g(z)}{F(z)}$ which is regular in $|z|<1$ and satisfies $S(0)>0$ and $0<|S(z)|\leqslant 1$; $|S(e^{i\theta})|=1$ (almost everywhere).

It shows that $-\log|S(z)|$ is a positive harmonic function and vanishes almost everywhere on the boundary of the unit disk.

It follows from the Herglotz representation (Theorem 2. 1. 1) that $-\log|S(z)|$ can be expressed as the Poisson-Stieltjes integral of a bounded nondecreasing function $\mu(t)$ with $\mu'(t)=0$ (almost everywhere). It is easily shown that

$$S(z)=exp\left\{-\int_0^{2\pi}\frac{e^{it}+z}{e^{it}-z}d\mu(t)\right\}. \tag{2.2.5}$$

Combining the above analyses we derive the factorization:

$$f(z)=e^{i\gamma}B(z)S(z)F(z)$$

and call the function

$$F(z)=e^{i\gamma}exp\left\{\frac{1}{2\pi}\int_0^{2\pi}\frac{e^{it}+z}{e^{it}-z}\ log\ \psi(t)dt\right\} \tag{2.2.6}$$

an outer function for a H^p function, where γ is a real number,

$$\psi(t)\geqslant 0,\ log\ \psi(t)\in L^1 \text{ and } \psi(t)\in L^p.$$

Expression (2.2.4) represents an outer-function.

We call a function $f(z)$ regular in $|z|<1$ an inner function, if it satisfies

$$|f(z)|\leqslant 1,$$

$$|f(e^{i\theta})|=1 \text{ (almost everywhere)}.$$

We know that every inner function has a factorization of the form $e^{i\gamma}B(z)S(z)$, where $B(z)$ is the Blaschke product, $S(z)$ assumes the form of (2.2.5), with $\mu(t)$ being a bounded nondecreasing singular function and $\mu'(t)=0$ almost everywhere. Functions like $S(z)$ are called singular inner functions.

Theorem 2.2.8. (Normal factorization theorem). *Every function $f(z)\in H^p$ with $f(z)\not\equiv 0$ has a unique factorization of the form:*

$$f(z)=B(z)S(z)F(z),$$

where $B(z)$ is a Blaschke product, $S(z)$ is a singualr inner function, and $F(z)$ is an outer function. Conversely, a product of such functions $B(z)S(z)F(z)\in H^p$.

Proof. From the previous discussion, the first part of the assertions follows. We only need to prove the converse part. To this end we shall only show that any outer function (2.2.6) must belong to H^p. By the well-known arithmetic mean and geometric mean inequality, we have

$$|F(z)|^p\leqslant\frac{1}{2\pi}\int_0^{2\pi}P(r,\theta-t)[\psi(t)]^p dt①.$$

① Arithmetic mean and geometric mean inequality: if $f(x)\geqslant 0$ and is integrable with respect to a non-negative measure dμ and $A(f)=\int f(x)d\mu(x)$ and $B(f)=exp\{A(log f)\}$ be the arithmetic mean and geometric mean of f, respectively. Then $B(f)\leqslant A(f)$; the "=" sign holds iff $f(x)\equiv$ constant.

Hence

$$\int_0^{2\pi} |F(re^{i\theta})|^p d\theta \leqslant \int_0^{2\pi} [\psi(t)]^p dt.$$

The theorem is thus proved.

Similar factorization holds for a function $f\in N$. In this case, the function $F(z)$ in expression (2.2.6) satisfying $\psi(t)\geqslant 0$, $\log \psi(t)\in L^1$, and we shall call such $F(z)$ an outer function of class N.

Theorem 2.2.9. *Every function $f\in N$ with $f\not\equiv 0$ can be represented in the form:*

$$f(z)=B(z)\left[\frac{S_1(z)}{S_2(z)}\right]F(z), \tag{2.2.7}$$

where $B(z)$ is a Blaschke product, $S_1(z)$ and $S_2(z)$ are singular inner functions, and $F(z)$ (with $(\psi(t)=|f(e^{it})|)$ is an outer function of class N. Conversely, every function having the form (2.2.7) must belong to N.

Proof. Assume $f(z)=e^{i\gamma}B(z)g(z)$ $(|z|<1)$, where $g\in N$ with $g(z)\neq 0$, $g(0)>0$. Since $\log|g(z)|\in h^1$, it can be represented by a Poisson-Stieltjes integral. Namely

$$\log|g(z)|=\frac{1}{2\pi}\int_0^{2\pi} P(r,\theta-t)d\nu(t),$$

where $\nu(t)$ is of bounded variation. By decomposing $\nu(t)$ into the absolutely continuous and singular parts, the assertion follows. The converse part is an immediate consequence of the observation that every Poisson-Stieltjes integral belongs to h^1.

3. H^p Is a Banach Space

3.1. POISSON INTEGRAL AND H^1

Theorem 2.3.1. *Let $f(z)$ be a regular function in $|z|<1$. Then a necessary and sufficient condition for f to be expressed as*

$$f(z)=\frac{1}{2\pi}\int_0^{2\pi} P(r,\theta-t)\varphi(t)dt \tag{2.3.1}$$

with $\varphi\in L^1$, is that $f\in H^1$ and $\varphi(t)=f(e^{it})$ (almost everywhere).

Proof. If a regular function f assumes form (2.3.1), then

$$\int_0^{2\pi} |f(re^{i\theta})|d\theta \leqslant \int_0^{2\pi} |\varphi(t)|dt,$$

hence $f \in H^1$.

Conversely, assume $f \in H^1$ and set

$$\Phi(z) = \frac{1}{2\pi}\int_0^{2\pi} P(r,\theta-t)f(e^{it})dt.$$

We have, to any ρ with $0<\rho<1$,

$$f(\rho z) = \frac{1}{2\pi}\int_0^{2\pi} P(r,\theta-t)f(\rho e^{i\theta})dt.$$

However, by Theorem 2. 2. 6,

$$\int_0^{2\pi} |f(\rho e^{it}) - f(e^{it})|dt \to 0 \qquad (\rho \to 1),$$

hence

$$f(\rho z) \to \Phi(z) \qquad (\rho \to 1).$$

It follows that $\Phi(z) = f(z)$. This also proves the theorem.

Theorem 2. 3. 2 . *If $f(z)$ is a function regular inside the unit disk with its real part being nonnegative, then f must belong to $H^p(p<1)$.*

Proof. We may assume without any loss of generality that $f(0)=1$. From the hypothesis we see that the range of $f(z)$ lies on the right half-plane. Therefore, f is subordinate to

$$\frac{1+z}{1-z} = P(r,\theta) + iQ(r,\theta),$$

where $P(r,\theta)$ is the Poisson kernel and

$$Q(r,\theta) = \frac{2r\sin\theta}{1-2r\cos\theta+r^2}$$

is the conjugate Poisson kernel.

Since $\frac{1+z}{1-z} \in H^p$, $P(1,\theta)=0$ $(\theta \neq 0)$, it follows from Theorem 2. 1. 7 that

$$\int_0^{2\pi} |f(re^{i\theta})|^p d\theta \leqslant \int_0^{2\pi} \left|\frac{1+re^{i\theta}}{1-re^{i\theta}}\right|^p d\theta$$

$$\leqslant \int_0^{2\pi} |Q(1,\theta)|^p d\theta < \infty (p<1).$$

The theorem is thus proved.

3. 2. BANACH SPACE

Lemma. *If* $f\in H^p(0<p<\infty)$, *then*

$$|f(z)|\leqslant 2^{\frac{1}{p}}\|f\|_p(1-r)^{-\frac{1}{p}},\ r=|z|.$$

Proof. We begin by noting that if $p\neq 1$, then the inequality can be derived easily by Theorem 2. 3. 1. For $p=1$, we may apply Theorem 2. 2. 5 to obtain $f=Bg$, where B is the Blaschke product and $g\in H^p$ with $g(z)\neq 0$. Thus $|f(z)|\leqslant|g(z)|$ and, hence, $\|f\|_p=\|g\|_p$. Since $g(z)\neq 0$, one of the branches of $[g(z)]^p$ is single-valued and regular in the unit disk and belongs to H^1. We now use the result for the case $p=1$ to derive the inequality.

Let

$$K^p=\{f(e^{i\theta}):f\in H^p\}.$$

Then $K^p\subset L^p$ obviously.

Theorem 2. 3. 3. K^p *is the closure (in the* L^p *norm,* $0<p<\infty$*) of the set of all the polynomials generated by* $e^{i\theta}$*; i. e., polynomials of the form:* $\sum_{k=0}^{n} a_k e^{ik\theta}$.

Proof. For $0<p<1$, we set $f_\rho(z)=f(\rho z)$. To any $f\in H^p$ and given $\varepsilon>0$, we can, by Theorem 2. 2. 6, choose $\rho<1$ such that

$$\|f_\rho-f\|_p<\varepsilon/2^{1+\frac{1}{p}}.$$

Let $S_n(z)$ denotes the partial sum of the first $n+1$ terms of the Taylor series expansion of f around $z=0$. Since $S_n(z)\to f(z)\ (|z|=\rho)$ uniformly, we have, for n sufficiently large,

$$\|S_n-f_\rho\|_p<\varepsilon/2^{1+\frac{1}{p}}.$$

Now by applying the well-known inequality

$$(a+b)^p\leqslant 2^p(a^p+b^p),\ a>0,\ b>0,\ 0<p<\infty,$$

we obtain immediately the following inequality

$$\|S_n-f\|_p<\varepsilon\ (n \text{ sufficiently large}). \tag{2. 3. 2}$$

As to the case $p\geqslant 1$, one also can derive the conclusion (2. 2. 3) by applying Minkowski inequality. (2. 2. 3) shows clearly that the boundary values function $f(e^{i\theta})$ is the closure (in L^p norm) of polynomials generated by $e^{i\theta}$. Finally, we need to prove that K is a closed set. Let $\{f_n(e^{i\theta})\}$ be a sequence of functions from K, which converges, in the L^p norm, to a function $\varphi(z)\in L^p$. According to the lemma, $\{f_n(z)\}$ is uniformly bounded on $|z|\leqslant R<1$. Thus $\{f_n(z)\}$ forms a normal family and, hence, a subsequence $\{f_{n_k}(z)\}$ of $\{f_n(z)\}$ can be chosen which converges uniformly to $f(z)$ on $|z|\leqslant R<1$ with $f\in H^p$. Now to any given $\varepsilon>0$, choose N sufficiently large such that

$$\|f_n-f_m\|_p<\varepsilon \qquad (n,m\geqslant N).$$

Further, when $m\geqslant N$, $r<1$, we have

$$M_p(r,f-f_m)=\lim_{k\to\infty} M_p(r,f_{n_k}-f_m).$$

Let $r\to 1$, we obtain

$$\|f-f_m\|_p\to 0 \qquad (m\geqslant N).$$

Hence $\|f-f_n\|_p\to 0$ and, moreover,

$$\varphi(\theta)=f(e^{i\theta}) \text{ (almost everywhere).}$$

Corollary. *If* $1\leqslant p\leqslant\infty$, *then* H^p *is a Banach space.*

The validity of the corollary for $1\leqslant p<\infty$ is obvious. For the caes $p=\infty$, one can derive that K^∞ is closed by method of the proof of Theorem 2. 3. 3. It follows that H^∞ is also a Banach space.

We note that $\|\cdot\|_p$ is not a norm for $0<p<1$, however, H^p can be shown to be a complete metric space.

CHAPTER 3

Vector-Valued Analysis

1. Vector-Valued Functions

1.1. VECTOR-VALUED BOUNDED VARIATIONS

Let E be a Banach space.

We denote the real and complex number systems by $\mathscr{R}$ and $\mathscr{C}$ respectively, and the adjoint space of E by E', i.e. E' is the set of linear bounded functionals on E.

Definition 3.1.1. A vector-valued function $f(t)$ on the interval $[\alpha,\beta]$ to E is of

(1) weak bounded variation in $[\alpha,\beta]$ if $\varphi(f(t))$ is of numerical bounded variation for every $\varphi\in E'$,

(2) bounded variation if

$$sup\|\sum_i (f(\beta_i)-f(\alpha_i))\|<\infty$$

over every choice of a finite number of non-overlapping intervals (α_i,β_i) in $[\alpha,\beta]$ and

(3) strong bounded variation if

$$sup\sum_i \|f(\alpha_i)-f(\alpha_{i-1})\|<\infty,$$

where all possible finite partitions of $[\alpha,\beta]$ are allowed. The two suprema are know as the total and the strong total variations respectively.

It is easy to verify that strong bounded variation implies bounded variation and that bounded variation implies weak bounded variation.

Theorem 3.1.1. *A function $f(t)$ of weak bounded variation is of bounded variation (but not neces-*

sarily of strong bounded variation).

Proof. *Var* $\{\varphi(f(t))\}$ is the total variation of $\varphi(f(t))$ on $[\alpha,\beta]$. Then for every choice of a finite number of non overlapping intervals (α_i,β_i) in $[\alpha,\beta]$ we have

$$\begin{aligned}&|\varphi(\{\sum_i(\beta_i)-f(\alpha_i)\})|\\ &\leqslant\sum_i|\varphi(f(\beta_i))-\varphi(f(\alpha_i))|\\ &\leqslant Var\{\varphi(f(t))\}.\end{aligned}$$

By the uniform boundedness theorem there exists an $M>0$ such that

$$\|\sum_i(f(\beta_i)-f(\alpha_i))\|\leqslant M$$

for all choices of a finite number of non-overlapping intervals in $[\alpha,\beta]$.

The second part of the theorem is demonstrated by means of a counter-example. Let E be the space of all bounded complex-valued functions $g(\tau)$ on the interval $[0,1]$ with

$$\|g\|=sup\{|g(\tau)|:0\leqslant\tau\leqslant1\}.$$

We next define the vector function $f(t)=g_t(\tau)$ on $[0,1]$ to E so that

$$g_t(\tau)=\begin{cases}1 & \text{for} \quad 0\leqslant\tau\leqslant t,\\ 0 & \text{for} \quad t\leqslant\tau\leqslant1,\end{cases}$$

$g_1(\tau)\equiv1$. For any choice of non-overlapping intervals (α_i,β_i) in $[0,1]$ we have $\|\sum_i f(\beta_i)-f(\alpha_i))\|\leqslant1$. On the other hand, $\|f(t_1)-f(t_2)\|=1$ for any choice of $t_1,t_2\in[0,1]$ with $t_1\neq t_2$; it is apparent that $f(t)$ is not of strong bounded variation.

1. 2. VECTOR-VALUED INTEGRATION

Let $\mathscr{R}=(-\infty,\infty)$ and $\mathscr{C}$ be a complex plane.

Definition 3. 1. 2. A vector function $f(t)$ defined on a subset S of $\mathscr{R}$ (or $\mathscr{C}$) with values in E is

(1) weakly continuous at $t=t_0$ if

$$\lim_{t\to t_0}|\varphi(f(t)-f(t_0))|=0 \text{ for each } \varphi\in E',$$

(2) strongly continuous at $t=t_0$ if

$$\lim_{t\to t_0}\|f(t)-f(t_0)\|$$

The Riemann-Stieltjes integral can be extended to vector-valued functions. Let $f(t)$ be a vector-valued function on the interval $[\alpha,\beta]$ to E and let $g(t)$ be a numerically-valued function on the same interval.

We denote the subdivision

$$\sigma_0=\alpha<t_1<\cdots<t_n=\beta$$

together with the point $\tau_i(t_{i-1}\leqslant\tau_i\leqslant t_i)$ by π and set

$$\|\pi\|=max\{|t_i-t_{i-1}|: i=1,\cdots,n\}.$$

Definition 3.1.3. Let

$$s_\pi(f,g)=\sum_{i=1}^{n}f(\tau_i)(g(t_i)-g(t_{i-1})). \tag{3.1.1}$$

Then if $\lim_{\|\pi\|\to 0} s_\pi$ exists in a given topology, we define this limit to be the integral

$$\int_\alpha^\beta f(t)dg(t) \tag{3.1.2}$$

relative to this topology.

Definition 3.1.4. Let

$$S_\pi(g,f)=\sum_{i=1}^{n}g(\tau_i)(f(t_i)-f(t_{i-1})).$$

Then if $\lim_{\|\pi\|\to 0} S_\pi(g,f)$ exists in a given topology, we define this limit to be the integral

$$\int_\alpha^\beta g(t)df(t) \tag{3.1.3}$$

relative to this topology.

Theorem 3.1.2. *If either of the integrals (3.1.2) or (3.1.3) exists in given topology, then both will exist in this topology and*

$$\int_\alpha^\beta f(t)dg(t)=f(t)g(t)\,|_\alpha^\beta-\int_\alpha^\beta g(t)df(t). \tag{3.1.4}$$

Proof. This is an immediate consequence of the following identity and its dual:

$$\sum_{i=1}^{n}f(t_i)(g(t_i)-g(t_{i-1}))$$

$$=f(\beta)g(\beta)-f(\alpha)g(\alpha)-\sum_{i=0}^{n}g(t_i)(f(\tau_{i+1})-f(\tau_i)),$$

where $\tau_0=\alpha$ and $\tau_{n+1}=\beta$. For it is clear that $(\tau_0=\alpha\leqslant\tau_1\leqslant\cdots\leqslant\tau_{n+1}=\beta)$ is likewise a subdivision of $[\alpha,\beta]$, that $\tau_i \leqslant t_i \leqslant \tau_{i+1}$, and also that $\max_i |\tau_{i+1}-\tau_i| \leqslant 2\|\pi\|$.

We shall therefore concern ourselves only with the existence of one or the other of the two integrals (3. 1. 2) and (3. 1. 3).

Theorem 3. 1. 3. *Suppose that either* (1) $f(t)$ *is a strongly continuous vector-valued function on* $[\alpha,\beta]$ *with values in E and* $g(t)$ *is a numerically-valued function of bounded variation on* $[\alpha,\beta]$, *or* (2) $f(t)$ *is a vector-valued function on* $[\alpha,\beta]$ *with values in E of bounded variation and* $g(t)$ *is a continuous numerically-valued function on* $[\alpha,\beta]$. *Then the integrals*

$$\int_\alpha^\beta f(t)dg(t) \text{ and } \int_\alpha^\beta g(t)df(t)$$

exist in the norm topology. Further, if $T: E \to E$ *is a closed linear operator on E to a Banach space F, if* $f(t) \in D(T)$, *where* $D(T)$ *is the domain of T, and* $T[f(t)]$ *is strongly continuous in case* (1) *or of bounded variation in case* (2), *then*

$$T\left(\int_\alpha^\beta f(t)dg(t)\right) = \int_\alpha^\beta T(f(t))dg(t) \tag{3. 1. 5}$$

and

$$T\left(\int_\alpha^\beta g(t)df(t)\right) = \int_\alpha^\beta g(t)d\,T(f(t)). \tag{3. 1. 6}$$

Proof. For case (1) it is clear that $f(t)$ is uniformly continuous in the norm topology on $[\alpha,\beta]$. Hence given $\varepsilon>0$ there exists an $\delta>0$ such that $\|f(t')-f(t'')\|<\varepsilon$ if only $|t'-t''|<\delta$. Consequently if $\|\pi_1\|, \|\pi_2\| < \frac{1}{2}\delta$ a simple calculation shows that

$$\|S_{\pi_1}-S_{\pi_2}\| \leqslant 2\varepsilon Var\{g(t)\}.$$

This proves the existence of the strong integral for case (1). For case (2), $g(t)$ is uniformly continuous on $[\alpha,\beta]$ and, choosing ε and δ in the same manner as above, we obtain

$$|\varphi(S_{\pi_1}-S_{\pi_2})| \leqslant 2\varepsilon Var\{\varphi(f(t))\}$$

for $\|\pi_1\|, \|\pi_2\| < \frac{1}{2}\delta$. Now

$$Var\{\varphi(f(t))\} \leqslant Var\{Re(\Phi(f(t))\} + Var\{Im(\varphi(f(t))\}$$
$$\leqslant 4 \sup |\varphi(\sum_i (f(\beta_i)-f(\alpha_i))|$$

with the summation taken over all finite sets of non-overlapping intervals (α_i,β_i) in $[\alpha,\beta]$. Hence by Definition 3. 1. 1 there will exist an $M>0$ such that $Var\{\varphi(f(t)\} \leqslant M\|\varphi\|$.

As a consequence

$$\|S_{\pi_1}-S_{\pi_2}\|=\sup_{\|\varphi\|=1}|\varphi(S_{\pi_1}-S_{\pi_2})|\leqslant 2M\varepsilon.$$

This establishes the existence or the strong integral for case (2). Bacause of Theorem 3. 1. 2 both integrals will exist in each case. We turn now to the second part of the theorem. For any π we have $T[S_\pi(f,g)]=S_\pi(Tf,g)$ because of the linearity of T. Further we have just shown that

$$\lim_{\|\pi\|\to 0} S_\pi(f,g)=\int_\alpha^\beta f(t)dg(t)$$

and applying the above result to $T(f(t))$ instead of $f(t)$ we see that

$$\lim_{\|\pi\|\to 0} T(S_\pi(f,g))=\lim_{\|\pi\|\to 0} S_\pi(Tf,g)=\int_\alpha^\beta T(f(t))dg(t).$$

Since T is closed, it follows that $\int_\alpha^\beta f(t)dg(t)\in D(T)$ and (3. 1. 5) holds. Similarly, we can obtain (3. 1. 6).

Definition 3. 1. 5. Let E, F be topological spce and let $(E,\sum,\mu)$ be a finite measure space. A mapping $f: E\to F$ is said to be simple if there are disjoint sets $A_1,\cdots,A_k\in\sum$ and vectors $b_1,\cdots,b_k\in F$ such that

$$f(t)=\sum_{j=1}^{k}\chi_{A_j}(t)b_j \qquad \textit{for all } t\in E.$$

Then for each $A\in\sum$ we define

$$\int_A f d\mu=\sum_{j=1}^{k}\mu(A\cap A_j)b_j.$$

Lemma 3. 1. 1. *Let $(E,\sum,\mu)$ be a finite measure space, and let $f: E\to F$ be a simple mapping. Then for each $A\in\sum$ and $\varphi\in F'$ we have that:*

(1) $\varphi\left(\int_A f d\mu\right)=\int_A \varphi(f)d\mu,$

(2) $\left\|\int_A f d\mu\right\|\leqslant\int_A\|f\|d\mu.$

The proof of lemma 3. 1. 1 *is obvious.*

Definition 3. 1. 6. Let $(E,\sum,\mu)$ be a finite measure space.

(1) A mapping $f: E\to F$ is said to be measurable if there esists a sequence of simple

mappings $f_n: E \to F$ which converges to f almost everywhere.

(2) A measurable mapping $f: E \to F$ is said to be Bochner integrable if there exists a sequence of simple mappings $f_n: E \to F$ such that

$$\lim_{n\to\infty} \int_A \|f_n - f\| d\mu = 0.$$

In this case we define

$$\int_A f d\mu = \lim_{n\to\infty} \int_A f_n d\mu$$

for each $A \in \sum$.

Theorem 3. 1. 4. Let $(E, \sum, \mu)$ *be a finite measure space, and let* $f: E \to F$ *be a Bochner integrable mapping. Then*

(1) *The function* $\varphi(f): E \to \mathscr{C}$ *is integrable and*

$$\varphi\left(\int_A f d\mu\right) = \int_A \varphi(f) d\mu$$

for each $\varphi \in F'$ *and* $A \in \sum$.

(2) *The function* $\|f\|: E \to \mathscr{R}$ *is integrable and*

$$\left\| \int_A f d\mu \right\| \leqslant \int_A \|f\| d\mu$$

for each $A \in \sum$.

Lemma 3. 1. 1 guarantees that the Bochner integral $\int_A f\, d\mu$ is well defined. Indeed, on the one hand Lemma 3. 1. 1 implies that $\{\int_A f_n d\mu\}$ is a Cauchy sequence, and on the other hand Lemma 3. 1. 1 guarantees that the definition of $\int_A$ f dμ is independent of the choice of the sequence $\{f_n\}$. Finally, from Lemma 3. 1. 1 and the definition of the Bochner integral we can easily obtain the theorem.

Theorem 3. 1. 5. *If the sequence of Bochner integrable functions* $\{f_n\}$ *converges almost everywhere to a limit function* f *and if there exists a fixed function* F, *which is Lebesgue integrable, such that* $\|f_n(t)\| \leqslant F(t)$ *for all* n *and* t, *then* f *is Bochner integrable and*

$$\lim_{n\to\infty} \int_A f_n d\mu = \int_A f d\mu,$$

for each $A \in \sum$.

1. 3. VECTOR-VALUED HÖLDER CONDITIONS

Suppose that $f(t)$ is defined on a path C in the complex plane $\mathscr{C}$, with range in a Banach space E, and that f satisfies the condition

$$\|f(t_1)-f(t_2)\| \leqslant a|t_1-t_2|^{\mu} \tag{3.1.7}$$

for any $t_1, t_2 \in C$, where a is a constant and $0<\mu\leqslant 1$. We say that f satisfies a Hölder condition of order μ and the set of all functions is denoted by $VH(\mu)$.

Theorem 3. 1. 6. *$f\in VH(\mu)$ if and only if for any $\varphi\in E'$, $\varphi(f(t))\in H(\mu)$, where $H(\mu)$ is the set of numerical functions which satisfy* (3. 1. 7).

Proof. Suppose for any $\varphi\in E'$

$$|\varphi(f(t_1)-f(t_2))| \leqslant a|t_1-t_2|^{\mu}$$

holds. It follows that

$$\sup_{(t_1,t_2)\in C\times C} |\varphi(\frac{f(t_1)-f(t_2)}{|t_2-t_2|^{\mu}})| \leqslant a<\infty.$$

From the theorem of uniform boundedness we obtain

$$\sup\{\|\frac{f(t_1)-f(t_2)}{|t_1-t_2|^{\mu}}\| : (t_1,t_2)\in C\times C\}<\infty.$$

The converse is clear from the fact that

$$|\varphi(f(t_1)-f(t_2))| \leqslant \|\varphi\| \|f(t_1)-f(t_2)\|$$

for any $\varphi\in E'$.

Theorem 3. 1. 7. *If vector-valued functions $f(t)$ and $g(t)$ are defined on the path C with range in a Banach algebra E_0, and if $f\in VH(\mu)$ and $g\in VH(v)$, then $f(t)g(t)\in VH(\lambda)$, where $\lambda=\min(\mu,v)$.*

Proof. Since

$$\begin{aligned}
&\|f(t_1)g(t_1)-f(t_2)g(t_2)\| \\
&\leqslant \|f(t_1)g(t_1)-f(t_2)g(t_1)\| + \|f(t_2)g(t_1)-f(t_2)g(t_2)\| \\
&\leqslant \|f(t_1)-f(t_2)\| \max\{\|g(t)\| : t\in C\} \\
&+ \max\{\|f(t)\| : t\in C\} \|g(t_1)-g(t_2)\|,
\end{aligned}$$

the result of the Theorem 3. 1. 7 holds.

Definition 3. 1. 7. An element x is said to be regular if there is an element x^{-1}, called the inverse of x, such that $x^{-1}x = xx^{-1} = I$, where I is the unit element.

Lemma 3. 1. 2. *If E_0 is a Banach algebra with unit element I, then every element in the open sphere $\|x-I\|<1$ has a inverse x^{-1} and*

$$x^{-1} = I + \sum_{n=1}^{\infty} (I-x)^n. \tag{3.1.8}$$

Proof. Since $I + \sum_{n=1}^{\infty} (I-x)^n$ is absolutely convergent,

$$(I-(I-x))(I+\sum_{n=1}^{\infty}(I-x)^n) = (I-\sum_{n=1}^{\infty}(I-x)^n)(I-(I-x)) = I.$$

Thus (3. 1. 8) holds.

Lemma 3. 1. 3. *The regular elements form an open set in the Banach algebra E_0.*

Proof. We denote the set of regular element by S. We shall show for $x_0 \in S$ that all of the elements in the open sphere $\|x-x_0\|<\|x_0^{-1}\|^{-1}$ also belong to S. For such an x we have $\|I-xx_0^{-1}\| = \|(x_0-x)x_0^{-1}\|<1$. By the previous Lemma 3. 1. 2

$$(xx_0^{-1})^{-1} = I + \sum_{n=1}^{\infty}(I-xx_0^{-1})^n = I + \sum_{n=1}^{\infty}((x_0-x)x_0^{-1})^n.$$

It follows that

$$x^{-1} = ((xx_0^{-1})x_0)^{-1} = x_0^{-1}(xx_0^{-1})^{-1}$$

$$= x_0^{-1} + x_0^{-1}\sum_{n=1}^{\infty}((x_0-x)x_0^{-1})^n.$$

Moreover the series shows that

$$\|x^{-1}-x_0^{-1}\| \leqslant \|x_0^{-1}\|^2\|x-x_0\|(1-\|x-x_0\|\|x_0^{-1}\|). \tag{3.1.9}$$

Hence we have the following result.

Lemma 3. 1. 4. *The inverse x^{-1} is a continuous function of x in S.*

Theorem 3. 1. 8. *If the inverse f^{-1} of f in $VH(\mu)$ exists as $t \in C$, then $f^{-1} \in VH(\mu)$.*

Proof. Lemma 3. 1. 4 gives that $(f(t))^{-1}$ is a continuous function of $f(t)$, thus it is a continuous function of t. It follows that $(f(t))^{-1}$ is bounded. The inequality (3. 1. 9) yields

$$\|(f(t_1))^{-1}-(f(t_2))^{-1}\|\leqslant\frac{\|(f(t_2)^{-1}\|^2\|f(t_1)-f(t_2)\|}{1-\|f(t_1)-f(t_2)\|\|f(t_2)^{-1}\|}$$

$$\leqslant\frac{aa_1|t_1-t_2|^\mu}{1-a_1\|f(t_1)-f(t_2)\|}$$

$$<\frac{aa_1|t_1-t_2|^\mu}{1-a_1\varepsilon} \quad \text{as} \quad |t_1-t_2|<\sigma$$

where $a_1=sup\{\|(f(t))^{-1}\|:t\in C\}$ and there exists $\varepsilon>0$ such that $a_1\varepsilon<1/2$ as $|t_1-t_2|<\delta$.

1. 4. VECTOR-VALUED REGULAR FUNCTIONS

Definition 3. 1. 8. Let $f(t)$ be a vector function defined on a domain D in $\mathscr{R}$ (or $\mathscr{C}$) with values in E. If there is an element $f'(t)\in E$ such that the difference quotient $\frac{f(t+h)-f(t)}{h}$ tends weakly (strongly) to $f'(t)$ as $h\to 0$, we call $f'(t)$ the weak (strong) derivative of $f(t)$ at t. We also say that $f(t)$ is weakly (strongly) differentiable at t. If $f(t)$ is weakly (strongly) differentiable at any t in D. We call f weakly (strongly) differentiable in D.

Definition 3. 1. 9. $f(z)$ is said to be regular in D if $\varphi(f(z))$ is regular in Cauchy′ s sense for every $\varphi\in E'$, where range of $f(z)$ is in E. If a vector-valued function f is regular in $\mathscr{C}$, then f is called an entire function or said to be entire.

Lemma 3. 1. 5. *For any numerically valued regular function in the domain D and any compact subset S, there is a finite quantity $M(f,s)$ such that for every choice of z, $z+a$ and $z+\beta$ in S*

$$\left|\frac{1}{a-\beta}\left(\frac{1}{a}f(z+a)-f(z)\right)-\frac{1}{\beta}(f(z+\beta)-f(z))\right|\leqslant M(f,S).$$

Proof. Since

$$\frac{1}{a\beta}\cdot\frac{1}{t-z}+\frac{1}{a(a-\beta)(t-z-a)}-\frac{1}{\beta(a-\beta)(t-z-\beta)}$$

$$=\frac{1}{(t-z)(t-z-a)(t-z-\beta)},$$

$$\frac{1}{2\pi i}\int_C \frac{f(t)dt}{(t-z)(t-z-\alpha)(t-z-\beta)}=$$

$$\frac{1}{\alpha-\beta}\left\{\frac{1}{\alpha}(f(z+\alpha)-f(z))-\frac{1}{\beta}(f(z+\beta)-f(z))\right\},$$

where C is a closed path in D having a positive minimal distance d both from S and from the boundary of D, we have that

$$\left|\frac{1}{\alpha-\beta}\left(\frac{1}{\alpha}\left\{(f(z+\alpha)-f(z))-\frac{1}{\beta}(f(z+\beta)-f(z))\right\}\right|\right.$$

$$\leqslant\frac{1}{2\pi}\left|\int_C \frac{f(t)dt}{(t-z)(t-z-\alpha)(t-z-\beta)}\right|$$

$$\leqslant\frac{1}{2\pi}max\{|f(t)|;t\in C\}\cdot\frac{1}{d^3}\cdot Var(C).$$

where $Var(C)$ is the length of C.

Theorem 3. 1. 9. *If $f(z)$ is regular in D with range in E, then $f(z)$ is strongly continuous and strongly differentiable in D, uniformly with respect to z in any compact subset of D.*

Proof. We apply Lemma 3. 1. 5 to the function $\varphi(f(z))$. We write

$$f(z;\alpha,\beta)=\frac{1}{\alpha-\beta}\{\frac{1}{\alpha}(f(z+\alpha)-f(z))-\frac{1}{\beta}(f(z+\beta)-f(z))\}.$$

The Lemma 3. 1. 5 then asserts that

$$|\varphi(f(z;\alpha,\beta))|\leqslant M(\varphi,f,S)$$

for every choice of $z, z+\alpha$ and $z+\beta$ in S. By the theorem of uniform boundedness.

This implies the existence of a finite $M(f,s)$ such that

$$\|f(z;\alpha,\beta)\|\leqslant M(f,s).$$

If we take the limit as $\beta\rightarrow 0$, we obtain

$$\|\frac{1}{\alpha}(f(z+\alpha)-f(z))-f'(z)\|\leqslant|\alpha|M(f,s)$$

for all z and $z+\alpha$ in S. Q. E. D.

Theorem 3. 1. 10.(Cauchy). *If $f(z)$ is a regular vector-valued function on the domain D with values in the B-space E, then*

$$\int_C f(z)dz=\theta$$

for every simple closed rectifiable contour C in D such that the interior of C belongs to D.

Proof. For any linear bounded functional $\varphi \in E'$ we have

$$\varphi\left(\int_C f(z)dz\right) = \int_C \varphi(f(z))dz = 0,$$

hence

$$\int_C f(z)dz = \theta.$$

Theorem 3.1.11. (Cauchy integral formula). *Let $f(z)$ be a regular function on the domain D with values in E. Let C be a closed path in D, the interior of which is in D, and let z be such that $\arg(t-z)$ increases by 2π when t describes C (positive orientation). Then*

$$f^{(n)}(z) = \frac{n!}{2\pi i}\int_C \frac{f(t)dt}{(t-z)^{n+1}} \quad \text{for } n = 0, 1, 2, \cdots. \tag{3.1.10}$$

Proof. Since $\frac{d}{dz}\varphi(f(z)) = \varphi(\frac{d}{dz}f(z)), (\varphi(f(z)))^{(n)} = \varphi(f^{(n)}(z))$.

Theorem 3.1.12. (Cauchy-Hadamard). *Given the power series*

$$\sum_{n=0}^{\infty} a_n(z-z_0)^n, \quad a_n \in E. \tag{3.1.11}$$

Set $\frac{1}{\rho} = \limsup_{n\to\infty} \|a_n\|^{\frac{1}{n}}$. Then the series is absolutely convergent for $|z-z_0| < \rho$ and divergent for $|z-z_0| > \rho$. The series converges to a regular function on $|z-z_0| < \rho$ with values in E, the convergence being uniform in every circle of radius less than ρ.

Proof. The classical argument for the convergence of the series applies if we replace absolute values by norms throughout. It remains to show that the function $f(z) \equiv \sum_n a_n(z-z_0)^n$ is regular for $|z-z_0| < \rho$. Since the series converges in norm in the disc, we have

$$\varphi(f(z)) = \sum_{n=0}^{\infty} \varphi(a_n)(z-z_0)^n$$

for each $\varphi \in E'$ and all $|z-z_0| < \rho$. The series on the right converges to a numerically-valued regular function in $|z-z_0| < \rho$ so that $f(z)$ is itself regular in the disc by Definition 3.1.9.

Theorem 3.1.13. *If $f(z)$ is a vector-valued regular function in $0 \leqslant R_1 < |z-z_0| < r_2 \leqslant \infty$,* *then*

$$f(z)=\sum_{-\infty}^{+\infty}a_n(z-z_0)^n \tag{3.1.12}$$

where $a_n=\frac{1}{2\pi i}\int_C f(t)(t-z_0)^{-n-1}dt$ *and* C, *for instance, is the circle* $|t-z_0|=r(R_1<r<R_2)$.

Proof.

$$\|a_n\|\leqslant\frac{1}{2\pi}\int_C\|f(t)\|\,|t-z_0|^{-n-1}|dt|$$

$$\leqslant M_r r^{-n},$$

where $M_r=\max\{\|f(t)\|:t\in C,\ C \text{ is the circle } |t-z_0|=r\}$. Setting $r=R_2-\varepsilon$, we get

$$\|\sum_{n=1}^{\infty}a_n|z-z_0\|\leqslant\sum_{n=1}^{\infty}\|a_n\|\cdot|z-z_0|^n$$

$$\leqslant M_{R_2-\varepsilon}\sum_{n=1}^{\infty}\frac{|z-z_0|^n}{(R_2-\varepsilon)^n},$$

for $|z-z_0|<R_2-\varepsilon$. Thus $\sum_{n=1}^{\infty}a_n(z-z_0)^n$ converges absolutely.

Similarly, $\sum_{n=-\infty}^{-1}a_n(z-z_0)$ $|z-z_0|>R_1+\varepsilon$ is also absolutely convergent.

It follows that $\sum_{n=-\infty}^{\infty}a_n(z-z_0)^n$ converges absolutely for $R_1+\varepsilon<|z-z_0|<R_2-\varepsilon$.

For any $\varphi\in E'$ we have

$$\varphi(\sum_{n=-\infty}^{\infty}a_n(z-z_0)^n)=\sum_{n=-\infty}^{\infty}\varphi(a_n)(z-z_0)^n,$$

while $\varphi(a_n)=\frac{1}{2\pi i}\int_C\varphi(f(t))(t-z_0)^{-n-1}dt$. Hence

$$\varphi(f(z))=\sum_{n=-\infty}^{\infty}\varphi(a_n)(z-z_0)^n.$$

Since φ is arbitrary (3.1.12) holds. Q.E.D.

Uniqueness theorem. *If* $f(z)$ *and* $g(z)$ *are regular in* D *with values in* E *and if* $f(z_n)=g(z_n)$, $n=1,2,\cdots$, *the points* $\{z_n\}$ *having a limit point in* D, *then* $f(z)=g(z)$ *in* D.

Proof. For any $\varphi\in E'$ by the classical uniqueness theorem we get

$$\varphi(f(z))\equiv\varphi(g(z)) \text{ as } z\in D.$$

therefore

$$f(z)\equiv g(z) \quad \text{as} \quad z\in D \qquad \text{Q. E. D.}$$

By a method similar to the classical method we can obtain the following result.

The maximum principle. *Let $f(z)$ be defined in a domain D of the extended plane and on its boundary C, regular in D and strongly continuous in $D\cup C$. If $\sup\{\|f(z)\|: z\in C\}=M$, then either $\|f(z)\|\equiv M$ or $\|f(z)\|<M$ in D.*

Liouville's theorem. *If a vector-valued function $f(z)$ is regular in $\mathscr{C}$, and if $\|f(z)\|$ is bounded in $\mathscr{C}$, then*

$$f(z)\equiv \textit{constant element}.$$

Proof. For any $\varphi\in E'$ we have

$$|\varphi(f(z))|\leqslant\|\varphi\|\|f(z)\|.$$

It follows that $\varphi(f(z))$ is bounded in $\mathscr{C}$. The classical Liouville Theorem yields $\varphi(f(z))\equiv$constant. If $f(z)\not\equiv$constant element, then there are points z_1, z_2 such that $f(z_1)=f_1, f(z_2)=f_2$, where $f_1\neq f_2$. Hence there is a $\varphi_0\in E'$ such that

$$\varphi_0(f(z_1))\neq\varphi_0(f(z_2))$$

which is a contradiction. Q. E. D.

Definition 3. 1. 10. Let f have an isolated singularity at $z=a$ and let

$$f(z)=\sum_{n=-\infty}^{\infty} a_n(z-a)^n$$

be its Laurent Expansion about $z=a$. Then the residue of f at $z=a$ is the coefficient a_{-1}. Denote this by $res\, f(a)$.

Methods of proof of the following theorems are the same as in complex analysis.

Theorem 3. 1. 14. *Let f be a vector-valued regular function except for a finite number of points $\{z_j\}_{j=1}^{n}$ in the domain in $\mathscr{C}$. Let C be a closed path contained in $D-\{z_j\}_1^n$. Then*

$$\frac{1}{2\pi i}\int_C f(z)dz=\sum_{j=1}^{n} n(C,z_j)res f(z_j).$$

where $n(C,z_j)=\dfrac{1}{2\pi i}\displaystyle\int_C \frac{dz}{z-z_j}$ is the topological index of $\mathscr{C}$ with respect ot z_j.

Theorem 3.1.15. (1) *If f has a pole of order 1 at z_0 then* $res\ f(z_0)=\lim_{z\to z_0}(z-z_0)f(z)$.

(2) *If f has a pole of order n at z_0, then*

$$res\ f(z_0)=\frac{1}{(n-1)!}\lim_{z\to z_0}\frac{d^{n-1}}{dz^{n-1}}((z-z_0)^n f(z)).$$

Theorem 3.1.16. *If $f(z)=a_m(z-z_0)^m+a_{m+1}(z-z_0)^{m+1}+\cdots$, where $a_k\in E$ for $k=m,m+1,\cdots$, and if a_m^{-1} exists, then $f^{-1}(z)$ exists and has a pole with order m at z_0.*

Proof. Since

$$\frac{f(z)}{(z-z_0)^m}=a_m+a_{m+1}(z-z_0)+a_{m+2}(z-z_0)^2+\cdots,$$

$$\left\|\frac{f(z)}{(z-z_0)^m}-a_m\right\|<\|a_m^{-1}\|^{-1}\quad \text{as } 0<|z-z_0|<\delta,$$

where δ is sufficiently small. Thus

$$\left\|I-\frac{f(z)}{(z-z_0)^m}\ a_m^{-1}\right\|=\left\|(a_m-\frac{f(z)}{(z-z_0)^m})a_m^{-1}\right\|<1.$$

It follows that

$$\left(\frac{f(z)}{(z-z_0)^m}\ a_m^{-1}\right)^{-1}=I+\sum_{n=1}^{\infty}\left(I-\frac{f(z)}{(z-z_0)^m}\ a_m^{-1}\right)^n.$$

Therefore

$$f^{-1}(z)=\frac{a_m^{-1}}{(z-z_0)^m}+\frac{b_{m-1}}{(z-z_0)^{m-1}}+\frac{b_{m-2}}{(z-z_0)^{m-2}}+\cdots,$$

where $b_{m-1},b_{m-2},\cdots\in E$.

Corollary. *If $f(z)$ has a pole of order m at z_0, i,e.*

$$f(z)=\frac{a_{-m}}{(z-z_0)^m}+\frac{a_{-m+1}}{(z-z_0)^{m-1}}+\cdots,$$

and a_{-m}^{-1} exists, then $f^{-1}(z)$ exists and has a zero of order m at z_0.

We obtain this result using the methods of Theorem 3.1.16.

1.5. COMPACTNESS AND COVERGENÇE IN THE SPACE OF VECTOR-VALUED REGULAR FUNCTIONS

Suppose that D is a domain in $\mathscr{C}$, and that $\{f(z)\}$ is the set of vector-valued regular functions with range in E, where $z\in D$. Further suppose S is any compact set in D.

Definition 3. 1. 11. For any compact set S in D if an arbitrary sequence $\{f_n\}(\subset\{f\})$ has a strongly uniformly convergent subsequence $\{f_{n_k}\}$ on S, $\{f\}$ is said to be normal in D.

If the set $\{f\}$ of numerically valued fanctions is uniformly bounded in any compact S in D, then $\{f\}$ is normal in D. But in the vector-valued case the result may not hold. For example, if D is the unit disc, we set

$$f_1(z)=(z,0,\cdots),$$
$$f_2(z)=(0,z,0,\cdots),$$
$$f_3(z)=(0,0,z,0,\cdots),$$
$$\vdots$$

with range in l_2, then $\|f_n(z)\|=|z|<1$ for $n=1,2,\cdots$. But we have

$$\|f_n(z)-f_m(z)\|=\sqrt{2}\,|z| \qquad \text{as } |z|<1$$

for any integers $m\neq n$. Thus $\{f_n\}$ is not normal in D.

Theorem 3. 1. 17. *If $\{f\}$ is normal in D with range in E, then $\{f\}$ is uniformly bounded on any compact set S in D.*

Proof. For any $\varphi\in E'$ from the hypothesis we obtain $\{\varphi(f(z))\}$ is normal in D, hence $\{\varphi(f(z))\}$ is uniformly bounded on any compact set $S\subset D$. Therefore $\{f\}$ is uniformly bounded by the uniform boundedness theorem.

Theorem 3. 1. 18. *If the set $\{f\}$ of vector-valued regular functions is uniformly bounded on any compact set $S\subset D$, and if D is a simply connected domain, then $\{f\}$ is equicontinuous on S.*

Proof. Suppose that z_0 is any point in D, and that the disc $S(z_0, 2r_0)$ with the center z_0 and the radius $2r_0$ is in D. There exists a constant $M>0$ such that

$$\|f(z)\|\leqslant M \text{ as } z\in S(z_0, 2r_0)$$

for any $f\in\{f\}$ by the uniform boundedness. From Theorem 3. 1. 11 we get

$$\|f'(z)\|=\left\|\frac{1}{2\pi i}\int_C\frac{f(t)}{(t-z)^2}dt\right\|\leqslant\frac{M}{2\pi}\cdot 4\pi r_0\cdot\frac{1}{r_0^2}=\frac{2M}{r_0}$$

for $|z-z_0|\leqslant r_0$, where C is the circle $|z-z_0|=2r_0$. Because S is compact, $\|f'(z)\|$

$\leqslant M_1(S)$ as $z \in S$, where $M_1(S)$ is a constant dependent on S. By the vector-valued Cauchy theorem we obtain

$$f(z_2) - f(z_1) = \int_{z_1}^{z_2} f'(z)dz,$$

hence

$$\|f(z_2) - f(z_1)\| \leqslant \int_{z_1}^{z_2} \|f'(z)\| |dz| \leqslant M_1 |z_2 - z_1|. \qquad \text{Q. E. D.}$$

Theorem 3.1.19. *The set $\{f\}$ of vector-valued regular functions is normal in a simply connected domain D if and only if the range of $\{f\}$ in any compact set $S \subset D$ is sequentially compact.*

Proof. Suppose the range A of $\{f\}$ in S is sequentially compact. There exists a set $\{z_n\}_1^\infty$ such that $\overline{\{z_n\}} = S$. For arbitrary $\{f_n\} \subset \{f\}$ and z_1 in S, there exists a convergent subsequence $\{f_{1,n}(z_1)\} \subset \{f_n(z_1)\}$. Generally, from $\{f_{k-1,n}(z)\}$ we can select a convergent subsequence $\{f_{k,n}(z_k)\}$; then $\{f_{n,k}\}$ is uniformly convergent on S. In fact, for any $\varepsilon > 0$ by the compactness of S we get

$$\bigcup_{j=1}^{r} S(z_{k_j}, \delta(\frac{1}{3}\varepsilon)) \supset S,$$

where $S(z_{k_j}, \delta(\frac{1}{3}\varepsilon))$ is the disc with the center z_{k_j} and the radius $\delta(\frac{1}{3}\varepsilon)$. Since $\{f_{n,n}(z_{k_j})\}$ is convergent,

$$\|f_{n,n}(z_{k_j}) - f_{m,m}(z_{k_j})\| < \frac{1}{3}\varepsilon \qquad \text{as } m > n > N_{k_j}.$$

For arbitrary $z \in S$ there exists $S(z_{k_j}, \delta(\frac{1}{3}\varepsilon))$ such that $z \in S(z_{k_l}, \delta(\frac{1}{3}\varepsilon))$. Since A is sequentially compact, A is bounded. It follows that $\{f\}$ is uniformly bounded on S. Thus $\{f\}$ is equicontinuous on S. We infer that

$$\|f_{n,n}(z) - f_{n,n}(z_{k_l})\| < \frac{1}{3}\varepsilon$$

and

$$\|f_{m,m}(z) - f_{m,m}(z_{k_l})\| < \frac{1}{3}\varepsilon$$

as $|z - z_{k_l}| < \delta(\frac{1}{3}\varepsilon)$ and $m > n > N = \max_{1 \leqslant j \leqslant r} N_{k_j}$. Therefore

$$\|f_{n,n}(z) - f_{m,m}(z)\| < \delta \text{ for } z \in S \text{ as } m < n < N,$$

i. e. $\{f_{n,n}\}$ is uniformly convergent on S.

Next suppose $\{f\}$ is normal in D. For any compact set $S\subset D$, any sequence $\{z_n\}\subset S$ and any $\{f_n\}\subset\{f\}$ we show that $\{f_n(z_n)\}$ has a convergent subsequence. In fact, from the completeness of the Banach space and the normality of $\{f\}$ we can choose $\{f_{n_k}\}\subset\{f_n\}$ such that $\{f_{n_k}\}$ uniformly converges to f_0 as $z\in S$ and $k\to\infty$.

On the other hand there exists $\{z_{n'_k}\}$ such that $z_{n'_k}\to z_0$ as $k\to\infty$, thus $z_0\in S$. Since $\{f_{n_k}\}$ is uniformly convergent,

$$\|f_{n'_k}(z_0)-f_0(z_0)\|<\frac{1}{2}\varepsilon \qquad \text{as } k>K_1,$$

and

$$\|f_{n'_k}(z_{n'_k})-f_{n'_k}(z_0)\|<\frac{1}{2}\varepsilon \qquad \text{as } k>K_2.$$

Thus

$$\|f_{n'_k}(z_{n'_k})-f_0(z_0)\|\leqslant\|f_{n'_k}(z_{n'_k})-f_{n'_k}(z_0)\|+\|f_{n'_k}(z_0)-f_0(z_0)\|$$
$$<\frac{1}{2}\varepsilon+\frac{1}{2}\varepsilon=\varepsilon$$

as $k>\max(K_1,K_2)=K$. Q. E. D.

1. 6. BOUNDARY PROPERTIES OF A CLASS OF VECTOR-VALUED REGULAR FUNCTIONS IN $S(0,1)$

Theorem 3. 1. 20. *Let $f(re^{i\theta})$ be a function on $S(0,1)$ with values in a complex reflexive Banach space E. If $\hat{F}(t)$ (i. e. $F(e^{it})$) is of strongly bounded variation on the interval $[-\pi,\pi]$, put*

$$f(z)=f(re^{i\theta})=\frac{1}{2\pi}\int_{-\pi}^{\pi}\frac{1+re^{i(\theta-t)}}{1-re^{i(\theta-t)}}d\hat{F}(t). \qquad (3.1.13)$$

Then $f(z)$ is a vector-valued regular funtion in $S(0,1)$.

Proof. By Theorem 3. 1. 3, the integral (3. 1. 13) is defined and $\hat{F}'(t)$ (strong derivation) exists almost everywhere (see [9]) and is Bochner integrable. From the boundary value theorem in Chapter 6 of [12] for any $\varphi\in E'$ we get

$$\varphi(f(z))=\frac{1}{2\pi}\int_{-\pi}^{\pi}\frac{1+re^{i(\theta-t)}}{1-re^{i(\theta-t)}}\varphi(\hat{F}'(t))dt$$

is regular, thus $f(z)$ is a vector-valued regular function.

Since E is reflexive, $\hat{F}'(t)$ exists almost everywhere; it is denoted by $g(t)$, i. e. $\hat{F}'(t)=g(t)$ almost everywhere.

Theorem 3.1.21. *If θ is a number such that the Bochner integral*

$$\int_{-\pi}^{\pi} \frac{g(\theta+t)-g(\theta-t)}{tg\,\frac{1}{2}t}\,dt \tag{3.1.14}$$

and $\hat{F}'(\theta)$ exist, then $\lim\limits_{r\to 1} f(re^{i\theta})$ exists and

$$\lim_{r\to 1} f(re^{i\theta}) = \hat{F}'(\theta) + i\hat{f}(\theta),$$

where

$$\hat{f}(\theta) = -\frac{1}{2\pi}\int_{-\pi}^{\pi} \frac{g(\theta+t)-g(\theta-t)}{tg\,\frac{1}{2}t}\,dt.$$

Proof. Let $H_r(\theta) = \dfrac{1+re^{i\theta}}{1-re^{i\theta}}$,

$$Re(H_r(\theta)) = \frac{1-r^2}{1-2r\cos\theta+r^2} = P_r(\theta),$$

and

$$Im(H_r(\theta)) = \frac{2r\sin\theta}{1-2r\cos\theta+r^2} = Q_r(\theta).$$

Put

$$f(r,\theta) = \frac{1}{2\pi}\int_{-\pi}^{\pi} g(\theta-t)Q_r(t)\,dt,$$

then

$$f(r,\theta) = \frac{1}{2\pi}\int_{-\pi}^{\pi} \frac{g(\theta+t)-g(\theta-t)}{2}Q_r(t)\,dt.$$

Set

$$\Phi_\theta(t) = \frac{g(\theta+t)-g(\theta-t)}{2tg\,\frac{1}{2}t},$$

then from (3.1.14) we get frist $\Phi_\theta(t)$ is Bochner integrable.

$$f(r,\theta) - \hat{f}(\theta) = \frac{1}{2\pi}\int_{-\pi}^{\pi} \Phi_\theta(t)\Big(1 - \frac{2r\sin t\; tg\,\frac{1}{2}t}{1-2r\cos t+r^2}\Big)dt$$

$$= \frac{1}{2\pi}\int_{-\pi}^{\pi} \Phi_\theta(t)\frac{(1-r)^2}{1-2r\cos t+r^2}\,dt. \tag{3.1.15}$$

Let

$$h_r(t)=\frac{(1-r)^2}{1-2r\cos t+r^2}.$$

Then $0<h_r(t)<1$ and $\lim\limits_{r\to 1} h_r(t)=0$ as $t\neq 0$. By Theorem 3. 1. 4 we get

$$\int_{-\pi}^{\pi}\Phi_\theta(t)h_r(t)dt\to 0 \quad \text{as } r\to 1.$$

i. e. $\lim\limits_{r\to 1} f(r,\theta)=\hat{f}(\theta)$.

On the other hand from Theorem 3. 1. 3 we obtain

$$\frac{1}{2\pi}\int_{-\pi}^{\pi}P'_r(\theta-t)\hat{F}(t)dt$$

$$=-\frac{1}{2\pi}P_r(\theta-t)\hat{F}(t)\,|_{-\pi}^{\pi}+\frac{1}{2\pi}\int_{-\pi}^{\pi}P_r(\theta-t)d\hat{F}(t)$$

$$=\frac{1}{2\pi}\int_{-\pi}^{\pi}P_r(\theta-t)\hat{F}(t).$$

Since

$$\frac{1}{2\pi}\int_{-\pi}^{\pi}P'_r(\theta-t)\hat{F}(t)dt=\frac{r}{2\pi}\int_{-\pi}^{\pi}K_r(t)\frac{\hat{F}(\theta+t)-\hat{F}(\theta-t)}{2\sin t}dt,$$

where $K_r(t)=-\frac{1}{r}\sin t\, P'_r(t)$ for $0<r<1$, $\{K_r(t)\}$ is an approximate identity (see p. 17 and p. 35 in [12]). ① Let

$$G_\theta(t)=\frac{\hat{F}(\theta+t)-\hat{F}(\theta-t)}{2\sin t},$$

then $\lim\limits_{t\to 0}\frac{\hat{F}(\theta+t)-\hat{F}(\theta-t)}{2\sin t}=F'(\theta)$ (strong limit).

Let $G_\theta(0)=\hat{F}'(\theta)$. Then $G_\theta(t)$ is strong continuous at $t=0$. It follows that

$$\left\|\frac{1}{2\pi}\int_{-\pi}^{\pi}K_r(t)G_\theta(t)dt-G_\theta(0)\right\|$$

$$\leqslant \sup_{\delta\leqslant|t|<\pi}\frac{1}{2\pi}K_r(t)\int_{[-\pi,\pi]-(-\delta,\delta)}\|G_\theta(t)-G_\theta(0)\|dt$$

$$+\frac{1}{2\pi}\int_{-\delta}^{\delta}K_r(t)\|G_\theta(t)-G_\theta(0)\|dt. \tag{3.1.16}$$

---

① Any sequence of Lebesgue-integrable functions K_n which possesses the following properties

(i) $K_n\geqslant 0$,

(ii) $\frac{1}{2\pi}\int_{-\pi}^{\pi}K_n(x)dx=1$,

(iii) If I is any open interval about $x=0$, then
$\lim\limits_{n\to\infty}\sup\limits_{x\notin I}|K_n(x)|=0 \qquad (|x|\leqslant\pi)$,
we call an approximate identity.

Since the right of (3. 1. 16) is arbitrarily small,

$$\lim_{r\to 1}\frac{1}{2\pi}\int_{-\pi}^{\pi}K_r(t)G_\theta(t)dt=\hat{F}'(\theta).$$

Because

$$\int_{-\pi}^{\pi}f(t)H_r(\theta-t)dt=\int_{-\pi}^{\pi}f(\theta-t)H_r(t)dt,$$

$$\lim_{r\to 1}f(re^{i\theta})=\hat{F}'(\theta)+i\hat{f}(\theta).$$ Q. E. D.

Theorem 3. 1. 22. *Suppose E is reflexive Banach space. If $f(re^{i\theta})$ is a bounded vector-valued regular function with range in E, where $z=re^{i\theta}\in S(0,1)$, then $f(e^{i\theta})$ exists almost everywhere.*

Proof. Let $\psi(z)=\int_0^z f(z)dz$ and let

$$M=\sup\{\|f(z)\|:z\in S(0,1)\}.$$

Then $\|\psi(z_1)-\psi(z_2)\|\leqslant M|z_1-z_2|$ for any $z_1,z_2\in\overline{S}(0,1)$. It follows that $\psi(z)$ is strongly continuous and of strongly bounded variation on $C(0,1)$, where $C(0,1)=\{z:|z|=1\}$. Since E is reflexive, $\psi'(z)$ exists almost everywhere on $C(0,1)$. For any $\varphi\in E'$ we have $\varphi(\psi')=\varphi(f)$ as $z\in C(0,1)$ almost everywhere, i. e. $\psi'(z)=f(z)$ on $C(0,1)$ almost everywhere.

1. 7. VECTOR-VALUED ELLIPTIC FUNCTIONS

Definition 3. 1. 12. If D is open in $\mathscr{C}$ and f is a vector-valued function on D which is regular except for poles, then f is called a vector-valued meromorphic function on D.

Definition 3. 1. 13. If f is a doubly-periodic vector-valued meromorphic function with periods ω_1 and ω_2, where $Im(\omega_1/\omega_2)\neq 0$, then f is called a vector-valued elliptic functions with periods ω_1 and ω_2 by *VEF*.

Obviously, the following results hold.

Theorem 3. 1. 23. *If $f\in VEF$, then the derivative $f\in VEF$.*

Proof. For arbitrary $\varphi\in E'$ we have

$$\varphi(f(z+\omega_1+\omega_2))'=\varphi(f'(z))$$

and

$$\varphi(f(z+\omega_1+\omega_2))' = \varphi(f'(z+\omega_1+\omega_2)).$$

Thus

$$f'(z+\omega_1+\omega_2)=f'(z).$$

Since singular points of f' and f are same, $f' \in VEF$. Q. E. D.

Theorem 3. 1. 24. *If $f \in VEF$, and f is not a constant element, then f has at least one pole.*

Proof. Let us suppose that f does not have any poles. Then f is entire. It follows that $\|f(z)\|$ is bounded on the fundamental parellelogram P_0, where

$$P_0=\{2(t_1-1)\omega_1/2+(2t_2-1)\omega_2/2 : 0\leqslant t_j<1, \text{for } j=1,2\}.$$

By the vector-valued Liouville Theorem we get that f is a constant element. We have obtained the desired contradiction. Thus f has at least one pole.

Corollary. *If two functions f_1 and $f_2 \in VEF$ have the same poles on the periodic parallelogram with the same principal part at poles, then they are same except for a constant element.*

Proof. f_1-f_2 is entire and is bounded by hypotheses, hence f_1-f_2 is a constant element.

Theorem 3. 1. 25. *If $f \in VEF$, and f^{-1} exists (i. e. $f \cdot f^{-1}=I$), then $f^{-1} \in VEF$.*

Proof. Since

$$\begin{aligned} f^{-1}(z)f(z)=f(z)f^{-1}(z)&=I \\ &=f^{-1}(z+m\omega_1+n\omega_2)f(z+m\omega_1+n\omega_2) \\ &=f(z+m\omega_1+n\omega_2)f^{-1}(z+m\omega_1+n\omega_2), \\ f^{-1}(z)&=f^{-1}(z+m\omega_1+n\omega_2), \end{aligned}$$

where m, n are any integers. Q. E. D,

Theorem 3. 1. 26. *If $f \in VEF$, then the sum of the residues of all poles on the periodic parallelogram equals the zero element θ.*

Proof. Suppose s is the sum of the residues of all poles on P_0, then $\varphi(s)=0$ for any φ in E'. Thus $s=\theta$.

Theorem 3.1.27. *Suppose that E is a commutative Banach algebra with unit element and Schauder basis $\{e_k\}$. Suppose further that $f,g\in VEF$ have the same zeros $\alpha_1,\cdots,\alpha_n$ and poles $\beta_1,\cdots,\beta_n$ with the same order s, and that either f^{-1} or g^{-1} exists. Then f,g are the same except for a constant element.*

Proof. If f and g are regular at z_0, then f, g can be expanded in power series in a neighborhood of z_0:

$$f(z)=\sum_{n=0}^{\infty}a_n(z-z_0)^n, \tag{3.1.17}$$

$$g(z)=\sum_{n=0}^{\infty}b_n(z-z_0)^n, \tag{3.1.18}$$

where $a_n,b_n\in E$ for $n=0,1,2,\cdots$. Let

$$a_n=\sum_{j=1}^{\infty}\alpha_j^{(n)}e_j,\qquad b_n=\sum_{j=1}^{\infty}\beta_j^{(n)}e_j,$$

where $\alpha_j^{(n)},\beta_j^{(n)}\in\mathscr{C}$ for $j=1,2,\cdots$ and $n=0,1,2,\cdots$. It follows that

$$f(z)=\sum_{j=1}^{\infty}\alpha_j^{(0)}e_j+\sum_{j=1}^{\infty}\alpha_j^{(1)}e_j(z-z_0)+\sum_{j=1}^{\infty}\alpha_j^{(2)}e_j(z-z_0)^2+\cdots.$$

Since a power series is absolutely and uniformly convergent in any $\overline{S}(z_0,r_1)$ where $0<r_1<r_f$ and r_f is the radius of convergence of (3.1.17), for any $\varepsilon>0$ there exists N such that for $n>N$

$$\|f(z)-\sum_{k=1}^{n}\sum_{j=1}^{\infty}\alpha_j^{(k)}e_j(z-z_0)^k\|<\varepsilon.$$

Because

$$\sum_{k=1}^{n}\sum_{j=1}^{\infty}\alpha_j^{(k)}e_j(z-z_0)^k=\sum_{j=1}^{\infty}\sum_{j=1}^{n}\alpha_j^{(k)}e_j(z-z_0)^k,$$

$$f(z)=\sum_{j=1}^{\infty}\sum_{k=1}^{\infty}\alpha_j^{(k)}(z-z_0)^k e_j=\sum_{j=a}^{\infty}f_j(z)e_j.$$

Similarly,

$$g(z)=\sum_{j=1}^{\infty}g_j(z)e_j,$$

where $f_j(z)$ and $g_j(z)$ are $\sum_{k=1}^{\infty}\alpha_j^{(k)}(z-z_0)^k$ and $\sum_{k=1}^{\infty}\beta_j^{(k)}(z-z_0)^k$ respectively, and both are numerical elliptic functions. In fact, by Theorem 2.7.5 in [11] we obtain that for each j there exists $\varphi_j\in E$; such that

$$\varphi_j(e_j)\neq 0, \qquad \varphi_j(L_{e_j})=0,$$

where L_{e_j} is the complementary subspace of e_j. Thus

$$\varphi_j(f(z))=\varphi_j(\sum_j f_j(z)e_j)=\varphi_j(e_j)f_j(z).$$

Since $\varphi_j(f(z))$ is a numerical elliptic function, so is $f_j(z)$.

Similarly, we can get that $g_j(z)$ is a numerical elliptic function. From Theorem 5.12 (in Chapter Ⅲ. 5 in [27]) we obtain

$$f_j(z)=s_j f_1(z) \quad \text{for } j=2,3,\cdots,$$

where s_j are constants. Therefore

$$F(z)=f_1(z)(e_1+\sum_{j=2}^{\infty} s_j e_j).$$

Similarly,

$$g(z)=g_1(z)(e_1+\sum_{j=2}^{\infty} t_j e_j),$$

where $g_1(z)$ is the numerical elliptic function, t_j are constants for $j=2,3,\cdots$.

Here without loss of generality, we assume f^{-1} exists. Since $f_1(z)=sg_1(z)$, where s is a constant,

$$g(z)=s^{-1}f(z)(e_1+\sum_{j=2}^{\infty} s_j e_j)^{-1}(e_1+\sum_{j=2}^{\infty} t_j e_j).$$

In a neighborhood of a pole the proof is similar. Q. E, D.

Definition 3.1.14. Suppose $f\in VEF$ and λ is an arbitrary element in the range of $f(z)$. If the number of roots, counted according to their multiplicities, of the equation $f(z)=\lambda$ is always s, f is called a vector-valued elliptic function of order s.

Theorem 3.1.28. *Suppose that E is a Banach space with Schauder basis $\{e_n\}_1^{\infty}$, and that $f\in VEF$ is of order s, with poles $\beta_1,\cdots,\beta_q$ and orders $s_1,\cdots,s_q$ respectively. Then*

$$f(z)=C+\sum_{k=1}^{q}(B_{k1}\zeta(z-\beta_k)+B_{k2}\wp(z-\beta_k)-\frac{B_{k3}}{2!}\wp'(z-\beta_k)$$
$$+\cdots+(-1)^{s_k}(B_{ks_k}/(s_k-1)!)\wp^{(s_k-2)}(z-\beta_k)),$$

where $\sum_{k=1}^{q} S_k=S$, $\sum_{k=1}^{q} B_{k1}=\theta$, $C, B_{kj}\in E$ for $k=1,2,\cdots,q$ and $j=1,2,\cdots$, $\zeta(z)$, $\wp(z)$ are Weierstrass' s functions (see Chapter Ⅲ. 5 in [27]).

Proof. From Theorem 3. 7. 5, we get

$$f(z)=\sum_{n=1}^{\infty}f_n(z)e_n$$
$$=\sum_{n=1}^{\infty}\Big(c_n+\sum_{k=1}^{q}\Big(b_{k1}^{(n)}\zeta(z-\beta_k)+b_{k2}^{(n)}\mathscr{P}(z-\beta_k)-\frac{b_{k3}^{(n)}}{2!}\mathscr{P}'(z-\beta_k)$$
$$+\cdots+(-1)^{s_k}\frac{b_{ks_k}^{(n)}}{(s_k-1)!}\mathscr{P}^{(s_k-2)}(z-\beta_k)\Big)\Big)e_n.$$

Hence

$$f(z)=\sum_{n=1}^{\infty}c_ne_n+\sum_{k=1}^{q}\Big(\sum_{n=1}^{\infty}b_{k1}^{(n)}e_n\zeta(z-\beta_k)+\sum_{n=1}^{\infty}b_{k2}^{(n)}e_n\mathscr{P}(z-\beta_k)$$
$$+\cdots+(-1)^{s_k}\sum_{n=1}^{\infty}b_{ks_k}^{(n)}\mathscr{P}^{s_k-2}(z-\beta_k)e_n\Big)$$
$$=C+\sum_{k=1}^{q}\Big(B_{k1}\zeta(z-\beta_k)+\cdots+(-1)^{s_k}\frac{B_{ks_k}}{(s_k-1)!}\mathscr{P}^{(s_k-2)}(z-\beta_k),$$

where $C=\sum_{n=1}^{\infty}c_ne_n$, $B_{kj}=\sum_{n=1}^{\infty}b_{kj}^{(n)}e_n$ for $k=1,2,\cdots,q$ and $j=1,\cdots,s_k$.

2. Vector-Valued Boundary Value Problems

2. 1. VECTOR-VALUED CAUCHY TYPE INTEGRALS

Definition 3. 2. 1. Let C be a smooth curve without self-intersection. Suppose $t\in C$ and a circle $C(t_0,\varepsilon_0)$ is drawn with the center t_0 and the radius ε_0 such that there are only two intersection points t' and t'' on $C(t_0,\varepsilon)\cap C$. Considering the integral

$$\frac{1}{2\pi i}\int_{C-l_\varepsilon}\frac{f(t)}{t-t_0}dt,$$

where l_ε is the curve $C(t_0,\varepsilon)\cap C$. If

$$\lim_{\varepsilon\to 0}\frac{1}{2\pi i}\int_{C-l_\varepsilon}\frac{f(t)}{t-t_0}dt$$

exists, it is called the Cauchy principal value integral of $f(t)$.

Theorem 3. 2. 1. *Suppose that C is a path and $t_0\in C$, that $f(t)\in VEF(\mu)$ in a neighborhood of t_0, and that f is strongly continuous on C. Then*

$$\lim_{\varepsilon\to 0}\frac{1}{2\pi i}\int_{C-l_\varepsilon}\frac{f(t)}{t-t_0}dt$$

exists.

Proof. Obviously,

$$\frac{1}{2\pi i}\int_{C-l_\varepsilon}\frac{f(t)}{t-t_0}dt$$

exists. Let $C\cap C(t_0,\varepsilon')=l_{\varepsilon'}$ and suppose $l_{\varepsilon'}\subset l_\varepsilon$. We have

$$\left\|\int_{l_\varepsilon-l_{\varepsilon'}}\frac{f(t)-f(t_0)}{t-t_0}dt\right\|\leqslant\int_{l_\varepsilon-l_{\varepsilon'}}\frac{\|f(t)-f(t_0)\|}{|t-t_0|}|dt|$$

$$\leqslant\int_{l_\varepsilon-l_{\varepsilon'}}a|t-t_0|^{\mu-1}|dt|.$$

The remainder of the proof is easy by the completeness of the Banach space and standard methods used m the complex analysis. Q. E. D.

By the classical methods we can obtain the following result.

Theorem 3. 2. 2. *Suppose that $f\in VH(\mu)$ is a function of $t\in C$, where C is a closed path. Let $\Phi(z)=\frac{1}{2\pi i}\int_C\frac{f(t)}{t-z}dt$. Then*

$$\Phi^+(t_0)=\frac{1}{2\pi i}\int_C\frac{f(t)-f(t_0)}{t-t_0}dt+\frac{f(t_0)}{2\pi i}ln\frac{t_0-b}{t_0-a},$$
$$\Phi^-(t_0)=\frac{1}{2\pi i}\int_C\frac{f(t)-f(t_0)}{t-t_0}dt+\frac{f(t_0)}{2\pi i}ln\frac{t_0-b}{t_0-a}-f(t_0), \tag{3. 2. 1}$$

where $\Phi^+(t)$ and $\Phi^-(t)$ are the boundray values of $\Phi(z)$ as z tends to C from the interior and the exterior of C respectively.

The formula (3. 2. 1) is called the vector-valued Plèmelj formula.

Theorem 3. 2. 3. *If $f\in VH(\mu)$ as a function of $t\in C$, then $\Phi^\pm(t)\in VH(\mu)$ for $\mu<1$ and $f\in VH(\mu-\varepsilon)$ for $\mu=1$.*

Proof. From Theorem 3. 2. 2 we find that $\Phi^\pm(t)$ exists. For any $\varphi\in E'$, $\varphi(\Phi^\pm(t))\in H(\mu)$ for $\mu<1$, hence $\Phi^\pm(t)\in VH(\mu)$, where $H(\mu)$ is a set of numerical valued functions which satisfy the Hölder condition,

The proof of the case $\mu=1$ is similar.

Theorem 3. 2. 4. *If $f\in VH(\mu)$ on C, then*

$$\Phi(t)=\frac{1}{2\pi i}\int_C \frac{f(\tau)}{\tau-t}d\tau \in VH(\mu) \quad for\ t\in C.$$

The corresponding classical theorem yields Theorem 3. 2. 4.

Lemma 3. 2. 1. *Suppose that the vector-valued function $K(z,t)$ is regular for z in D^+, which is the interior of C for each fixed t on C, that $k(z,t)$ is continuous as a function of z in the closed domain $\overline{D}$ for each fixed t on C, and that for each fixed z in $\overline{D}$, $K\in VH(\mu)$. Then*

(1) $\Phi(z)=\frac{1}{2\pi i}\int_C \frac{K(z,t)}{t-z}dt$ *is regular when z is in D,*

(2) $\Phi(z)\in VH(\mu)$ *as* $z\in\overline{D}$,

(3) $\Phi^+(t)=\frac{1}{2}K(t,t)+\frac{1}{2\pi i}\int_C \frac{K(t,\tau)}{\tau-t}d\tau$ *for* $t\in L$.

Proof. By methods similar to those of complex analysis we can obtain that $\Phi^+(t)$ exists. For any $\varphi\in E'$ we have that

$$\varphi(\Phi(z))=\frac{1}{2\pi i}\int_C \frac{\varphi(K(z,t))}{t-z}dt$$

is regular in D, thus $\Phi(z)$ is a vector-valued regular function.

On the other hand,

$$(\varphi(\Phi(t)))^+=\varphi(\Phi^+(t))$$

$$=\frac{1}{2}\varphi(K(t,t))+\frac{1}{2\pi i}\int_C \frac{\varphi(K(t,\tau))}{\tau-t}d\tau \ (t\in C).$$

Therefore (3) holds.

(1) and (2) are obvious.

Note. For each $\varphi\in E'$, if $(\varphi(\Phi(t)))^+$ is defined, we may not get that $\Phi^+(t)$ is defined. For example, we take $\Phi(z)=(1,z,z^2,\cdots,z^n,\cdots)$ in l_2, where $z\in S(0,1)$. For any $\varphi=(a_0,a_1,\cdots,a_n,\cdots)\in l_2=l_2'$, we have $\varphi(\Phi(z))=\sum_{n=0}^{\infty}a_n z^n$. Hence $(\varphi(\Phi(e^{i\theta})))^+=\sum_{n=0}^{\infty}a_n e^{in\theta}$ is defined, but $\Phi^+(e^{i\theta})$ can not be defined.

2. 2. VECTOR-VALUED SINGULAR INTEGRAL EQUATIONS

In this section, we shall deal with vector-valued singular integral equations with range in a Banach algebra E_0.

$$A(t)y(t)+\frac{1}{\pi i}\int_{C}\frac{K(t,\tau)}{\tau-t}y(\tau)d\tau=f(t) \text{ as } t\in C, \tag{3.2.2}$$

where C consists of a finite number of arc-wise smooth closed curves which do not cross each other; D^+ is the inner domain which is on the left of C traced along a positive direction; $A(z)$ is a reguar vector-valued function which is on D^+ with values in E_0 and satisfies a vector-valued Hölder condition on $\overline{D}$. For each fixed $w\in\overline{D}$, $K(z,w)$ is a regular vector-valued function in D^+ and for each fixed $z\in D^+$, $K(z,w)$ is a regular vector-valued function in D^+ and satisfies a vector-valued Hölder condition for both z and w on $\overline{D}$. The function $f(t)$ satisfies a vector-valued Hölder condition on C.

Under the conditions just mentioned, we shall study the representation of the solution $y(t)$ of (3.2.2).

Let $B(z)=K(z,z)$, then $B(z)$ is regular in D^+. In fact, for arbitrary $\varphi\in E'$, $\varphi(B(z))=\varphi(K(z,z))$ is regular by Hartogs theorem (see[2]). ①Thus $B(z)$ is regular in D^+. Since $\varphi(B(z))\in H(\mu)$ on $\overline{D}$, $B(z)\in VH(\mu)$ by Theorem 3.1.6.

Suppose that $A(t)\pm B(t)\neq\theta$ for $t\in C$, that $A(t)+B(t)$ has zeros $\alpha_1,\cdots,\alpha_m$, with multiplicities λ_k for $k=1,\cdots,m$ respectively, that $A(z)-B(z)$ has zeros $\beta_1\cdots,\beta_n$ with multiplicities μ_j for $j=1,\cdots,n$, that $(A(t)\pm B(t))^{-1}$ exists for z in $\overline{D}$ except at the zeros of $A(z)\pm B(z)$, and that both of

$$\frac{d^{\mu_j}}{dz^{\mu_j}}(A(\beta_j)-B(\beta_j))$$

$$\frac{d\lambda_k}{dz^{\lambda_k}}(A(\alpha_k)+B(\alpha_k)) \tag{3.2.3}$$

have inverses.

Under the above conditions we obtain

$$y(t)=(A(t)+B(t))^{-1}A(t)(A(t)-B(t))^{-1}f(t)$$

---

① A function $F(z_1,z_2)$ defined in a region D of the space C_2 is said to be a regular analytic function if the partial derivatives $\frac{\partial f}{\partial z_1}$ and $\frac{\partial f}{\partial z_2}$ exist at every (interior) point of D.

F. Hartogs was the first to prove that if $F(z_1,z_2)$ is a regular analytic function in a bi-cylinder

$$|z_1|\leqslant r_1, \quad |z_2|\leqslant r_2 \tag{$*$}$$

then it can be expanded in a power series

$$\sum_{m,n=0}^{\infty}a_{mn}z_1^m z_2^n$$

that converges absolutely and uniformly in every closed subregion of ($*$).

$$-\frac{1}{\pi i}(A(t)+B(t))^{-1}\int_C \frac{K(t,\tau)}{\tau-t}(A(\tau)-B(\tau))^{-1}f(\tau)d\tau$$

$$+4(A(\tau)+B(\tau))^{-1}\sum_{j=1}^{n}\sum_{r=0}^{\mu_j-1}B_{jr}(t)C_{jr}, \tag{3.2.4}$$

where

$$B_{jr}(z)=\frac{1}{r!}\text{res}\left(\frac{K(z,\tau)(A(\tau)-B(\tau))^{-1}(\tau-\beta_j)^r}{\tau-z}\right)_{\tau=\beta_j} \tag{3.2.5}$$

for $r=0,1,\cdots,\mu_j-1$ and $j=1,\cdots,n$, *and* C_{jr} are defined by the following linear equations:

$$\sum_{j=1}^{n}\sum_{r=0}^{\mu_j-1}\frac{d^\sigma}{dz^\sigma}((A(z)-B(z))(A(z)+B(z))^{-1}B_{jr}(z))_{z=\alpha_k}\cdot C_{jr}$$

$$=\frac{1}{4\pi i}\int_C\frac{\partial^\sigma}{\partial z^\sigma}((A(z)-B(z))(A(z)+B(z))^{-1}K(z,\tau))_{z=\alpha_k}$$

$$(A(\tau)-B(\tau))^{-1}f(\tau)d\tau$$

for $\sigma=0,1,\cdots,\lambda_{k-1}$ and $k=1,\cdots,m$. (3.2.6)

The formula (3.2.6) has a solution exactly when (3.9.1) does.

We shall only give some of the main steps needed in the proof the result.

If equation (3.2.6) has solutions C_{jr}, then we can let

$$\tilde{y}(z)=(A(z)-B(z))(A(z)+B(z))^{-1}$$

$$\left\{\frac{1}{2\pi i}\int_C\frac{K(z,\tau)}{\tau-z}(A(\tau)-B(\tau))^{-1}f(\tau)d\tau-2\sum_{j=1}^{n}\sum_{r=0}^{\mu_j-1}B_{jr}(z)C_{jr}\right\}, \tag{3.2.7}$$

i. e. $\tilde{y}(z)$ is defined on $\overline{D}$. In fact, by (3.2.3) and Theorem 3.1.16 we can conclude that $(A(z)\pm B(z))^{-1}$ has poles of order μ_j at points α_k and β_j, and is regular except for points α_k and β_j. Because $(A(t)-B(t))^{-1}$exists on C, $(A(t)-B(t))^{-1}$, $K(z,t)(A(t)-B(t))^{-1}f(t)$ and $K(z,t)(A(t)-B(t))^{-1}(t-\beta_j)^r\in VH(\mu)$ for each fixed z on $D^+\cup C$. It is clear that the first term and second term in { } in (3.2.7) are regular. Thus $\tilde{y}(z)$ is defined for $z\in\overline{D}$. By Lemma 3.2.1 we show that $\tilde{y}^+(t)$ exists.

On the other hand from (3.2.3), Theorem 3.1.15 and Theorem 3.1.16, we get

$$B_{jr}(z)=\frac{1}{r!}res\left\{\frac{K(z,\tau)}{\tau-z}(A(\tau)-B(\tau))^{-1}(\tau-\beta_j)^r\right\}_{\tau=\beta_j}$$

$$=\frac{1}{r!(\mu_j-r-1)!}\frac{\partial^{\mu_j-r-1}}{\partial\tau^{\mu_j-r-1}}\left\{\frac{K(z,\tau)}{\tau-z}D_j^{-1}(\tau)\right\}_{\tau=\beta_j}$$

for $r=0,1,\cdots,\mu_{j-1}$ and $j=1,\cdots,n$, where

$D_j^{-1}(t)=(t-\beta_j)^{\mu_j}(A(t)-B(t))^{-1}$ and $C_{jr}=\tilde{y}^{(r)}(\beta_j)$.

Considering $\alpha_k \neq \beta_j$ and setting $z=\beta_j, j=1,\cdots,n$ in

$$\tilde{y}(z)=(A(z)-B(z))(A(z)+B(z))^{-1}$$

$$\left\{\frac{1}{2\pi i}\int_C \frac{K(z,t)}{t-z}(A(t)-B(t))^{-1}f(t)dt-2\sum_{j=1}^{n}\sum_{r=0}^{\mu_j-1}B_{jr}(z)\tilde{y}^{(r)}(\beta_j)\right\} \tag{3.2.8}$$

we see that $(A(z)+B(z))\tilde{y}(z)$ and $-2(A(z)-B(z))\sum_{r=0}^{\mu_j-1}B_{jr}(z)\tilde{y}^{(r)}(\beta_j)$ have identical expansions to order μ_j-1 at points $z=\beta_j$.

Letting

$$H_j(z)=B(z)D_j^{-1}(z),$$

we compare values between $F_1(z)=H_j(z)y(z)$ and

$$F_2(z)=(z-\beta_j)^{\mu_j}\sum_{r=0}^{\mu_j-1}\frac{1}{r!(\mu_j-r-1)!}\left(\frac{\partial^{\mu_j-r-1}}{\partial\tau^{\mu_j-r-1}}\frac{H_j(\tau)}{z-\tau}\right)_{\tau=\beta_j}\tilde{y}^{(r)}(\beta_j)$$

for the corresponding cases.

Since (3.2.3) holds, we get that $H_j(z)$ are regular in a neighborhood $U(\beta_j)$ of the points β_j for $j=1,\cdots,n$. From Theorem 3.2.2 we infer that

$$\frac{1}{2\pi i}\int_C \frac{K(z,\tau)}{\tau-z}(A(\tau)-B(\tau))^{-1}f(\tau)d\tau$$

is regular in D^+. From the condition (3.2.6) we can verify that $\tilde{y}(z)$ is regular in $U(\beta_j)$ for $j=1,\cdots,n$, as is $F_1(z)$.

Now the expansion to order μ_j-1 at $z=\beta_j$ of F_2 is

$$F_2(z)=H_j(\beta_j)\tilde{y}(\beta_j)+(H_j(z)\tilde{y}(z))_{\beta_j}(z-\beta_j)+\cdots$$
$$+(H_j(z)\tilde{y}(z))_{\beta_j}^{\mu_j-1}\frac{(z-\beta_j)^{\mu_j-1}}{(\mu_j-1)!}$$

Hence (3.2.8) holds.

Let $z\to t$, where $z\in D^+$ and $t\in L$, from Theorem 3.8.2 and

$$f(t)-[A(t)-B(t)][A(t)+B(t)]^{-1}B(t)[A(t)-B(t)]^{-1}f(t)$$
$$=[A(t)-B(t)][A(t)+B(t)]^{-1}A(t)[A(t)-B(t)]^{-1}f(t)$$

we can obtain through simplification

$$y(t)=[A(t)-B(t)]^{-1}[f(t)-2\tilde{y}^+(t)]. \tag{3.2.9}$$

Since $A(t)-B(t)$, $f(t)-2\tilde{y}^+(t)\in VH(\mu)$, by Theorem 3.3.2 and Theorem 3.3.3 we obtain that $y(t)\in VH(\mu)$. From (3.2.8) we infer

$$[A(z)+B(z)][A(z)-B(z)]^{-1}\tilde{y}(z)=\frac{1}{2\pi i}\int_C\frac{K(z,\tau)}{\tau-z}[A(\tau)-B(\tau)]^{-1}f(\tau)d\tau$$

$$-2\sum_{j=1}^{s}\sum_{r=0}^{\mu_j-1} res\left\{\frac{K(z,\tau)}{\tau-z}[A(\tau)-B(\tau)]^{-1}(\tau-\beta_j)^r\right\}_{\tau=\beta_j}\frac{\tilde{y}^{(r)}(\beta_j)}{r!}.$$

Using (3. 2. 9), Theorem 3. 1. 14 and Theorem 3. 1. 16 yield

$$\tilde{y}(z)=\frac{1}{2\pi i}\int_C\frac{K(z,t)}{t-z}[A(t)-B(t)]^{-1}f(t)dt$$

$$-\frac{1}{\pi i}\int_C\frac{K(z,t)}{t-z}[A(t)-B(t)]^{-1}\tilde{y}^+(t)dt$$

$$=\frac{1}{2\pi i}\int_C\frac{K(z,t)}{t-z}y(t)dt. \tag{3. 2. 10}$$

Now Lemma 3. 2. 1 gives

$$\tilde{y}^+(t)=\frac{1}{2}B(t)y(t)+\frac{1}{2\pi i}\int_C\frac{K(t,\tau)}{\tau-t}y(\tau)d\tau k. \tag{3. 2. 11}$$

When used in (3. 2. 9), (3. 2. 10) and (3. 2. 11) imply that $y(t)$ satisfies equation (3. 2. 1).

2. 3. VECTOR-VALUED DOUBLY-PERIODIC RIEMANN BAUNDARY VALUE PROBLEMS

VDF is the set of all vector-valued doubly-periodic continuous functions on P_0 (see Sec. 7) with periods ω_1 and ω_2, the range of which are in E.

Suppose an operator K is defined by the formula

$$Kf\equiv\frac{-b(t)}{\pi i}\int_C f(\tau)(\zeta(\tau-t)+\zeta(t))d\tau,$$

where $f(t),b(t)\in VDF\cap VH(\mu)$, and $\zeta(t)$ is the Weierstrass ζ-function.

Theorem 3. 2. 5. *If $f,b\in VDF\cap VH(\mu)$, then $Kf\in VDF\cap VH(\mu)$ and $\|Kf\|_\mu\leqslant q\|b\|_\mu\|f\|_\mu$, where q is a constant and* $\|f\|_\mu=\max_{t\in C}\|f(t)\|+\sup_{\substack{t_1,t_2\in C_0\\0<|t_1-t_2|\leqslant 1}}\frac{\|f(t_1)-f(t_2)\|}{|t_1-t_2|^\mu}$, *where C_0 is a path in P_0 which does not cross the point* 0.

Proof. Let $u(\tau-t)=\zeta(\tau-t)-\frac{1}{\tau-t}$. Then $u(\tau-t)$ is a continuous function of $\tau-t$.

Suppose that $\psi(t)=\frac{1}{\pi i}\int_C\frac{f(\tau)-f(t)}{\tau-t}d\tau$, and that s is the length of the arc from the starting point to the point t on C_0.

Suppose also that S_0, S are the lengths to the corresponding points t_0, t_1 respectively, that $t_1 - t_0 = h$ and $s_1 - s_0 = \sigma$. Finally suppose that the lengths of the arcs $\overset{\frown}{t' t_0}$ and $\overset{\frown}{t_0 t''}$ are 2σ, and that $l = \overset{\frown}{t' t_0} \cup \overset{\frown}{t_0 t''}$. We obtain

$$\Psi(t_1) - \Psi(t_0) = \frac{1}{\pi i} \int_{C_0} \frac{f(t) - f(t_0 + h)}{t - t_0 - h} dt - \frac{1}{\pi i} \int_{C_0} \frac{f(t) - f(t_0)}{t - t_0} dt$$

$$= \frac{1}{\pi i} \int_l + \frac{1}{\pi i} \int_{C_0 - l} = I_0 + I_1.$$

Then

$$\| f(t_1) - f(t_2) \| \leqslant \| f \|_\mu | t_1 - t_2 |^\mu,$$

$$\| I_0 \| \leqslant \frac{\| f \|_\mu}{\pi} \left(\int_l | t - t_0 - h |^{\mu - 1} ds + \int_l | t - t_0 |^{\mu - 1} ds \right)$$

$$\leqslant \frac{\| f \|_\mu}{\pi} k_0^{\mu - 1} \left(\int_{s_0 - 2\sigma}^{s_0 + 2\sigma} | s - s_0 - \sigma |^{\mu - 1} ds + \int_{s_0 - 2\sigma}^{s_0 + 2\sigma} | s - s_0 |^{\mu - 1} ds \right)$$

$$\leqslant \frac{\| f \|_\mu}{\pi \mu} k_0^{\mu - 1} (3^\mu + 2^{1 + \mu} + 1) \sigma^\mu$$

$$\leqslant \frac{\| f \|_\mu}{\pi \mu k_0} (3^\mu + 2^{1 + \mu} + 1) h^\mu$$

where k_0 is a constant which satisfies $0 < k_0 < \dfrac{| t_1 - t_2 |}{| s_1 - s_2 |} \leqslant 1$. Thus $\| f \| \leqslant m_1 \| f \|_\mu$, where m_1 is a constant.

We rewrite $I_1 = I' + I''$, where

$$I' = \frac{1}{\pi i} \int_{C_0 - l} \frac{f(t_0) - f(t_0 + h)}{t - t_0} dt,$$

$$I'' = \frac{1}{\pi i} \int_{C_0 - l} (f(t) - f(t_0 + h)) \left(\frac{1}{t - t_0 - h} - \frac{1}{t - t_0} \right) dt.$$

It follows that

$$\| I' \| = \left\| \frac{1}{\pi i} (f(t_0) - f(t_0 + h)) \log \frac{t' - t_0}{t'' - t_0} \right\|$$

$$\leqslant \frac{1}{\pi} \| f \|_\mu h^\mu \left| \arg \frac{t' - t_0}{t'' - t_0} \right|,$$

$$\leqslant \frac{\pi + \sigma_0}{\pi} \| f \|_\mu h^\mu$$

where σ_0 is a function of h, tending to 0 as $h \to 0$, hence there exists a constant m_2 such that

$$\| I' \|_\mu \leqslant m_2 \| f \|_\mu.$$

Similarly, we have

$$\|I''\|_\mu \leqslant m_3 \|f\|_\mu,$$

where m_3 is a constant.

Let $\psi_1(t) = \frac{1}{\pi i} \int_{C_0} f(\tau) u(\tau - t) d\tau$. Then ψ_1 is a regular function of t and

$$\psi_1'(t) = -\frac{1}{\pi i} \int_{C_0} f(\tau) u'(\tau - t) d\tau.$$

For arbitrary $\varepsilon > 0$ there exists δ $(0 < \delta < 1)$ such that

$$\frac{\|\psi_1(t) - \psi_1(t_0)\|}{|t - t_0|} \leqslant \|f\|_\mu \cdot \frac{1}{\pi} \max_{t \in C_0} \int_{C_0} \|u'(\tau - t)\| d\tau| + \varepsilon$$

as $|t - t_0| < \delta$ and $t \in C_0$, therefore $\psi_1 \in VH(1)$.

Since

$$\max_{t \in C_0} \|\psi_1(t)\| \leqslant \max_{t \in C_0} \|f(t)\| \cdot \max_{t \in C_0} \frac{1}{\pi} \int_{C_0} |u'(\tau - t)\| d\tau|,$$

hence

$$\|\psi_1\|_\mu \leqslant \|f\|_\mu (\max_{t \in C_0} \frac{1}{\pi} \int_{C_0} |u'(\tau - t)| \, |d\tau| + \max_{t \in C_0} \frac{1}{\pi} \int_{C_0} |u'(\tau - t)| \, |d\tau|).$$

Let

$$\psi_2(t) = \frac{1}{\pi i} \int_{C_0} f(\tau) \zeta(t) d\tau.$$

Then

$$\|\psi_2\|_\mu \leqslant \frac{1}{\pi} \|\zeta\|_\mu \cdot \|f\|_\mu Var(C_0),$$

where $Var(C_0)$ is the length of C_0.

Finally, we consider $\psi_3(t) = \frac{1}{\pi i} \int_{C_0} \frac{f(t)}{\tau - t} d\tau$, then clearly

$$\|\psi_3\|_\mu \leqslant \|f\|_\mu.$$

Combining the above we obtain

$$\|Kf\|_\mu \leqslant q \|f\|_\mu \cdot \|b\|_\mu,$$

where

$$q = \sum_{i=1}^{3} m_i + \max_{t \in C_0} \frac{1}{\pi} \int_{C_0} |u(\tau - t)| \, |d\tau|$$

$$+\max_{t\in C_0}\frac{1}{\pi}\int_{C_0}|u'(\tau-t)|\ |dt|+\frac{\|\zeta\|_\mu}{\pi}Var(C_0)+1.$$ Q. E. D.

We assume the operator R is defined by the formula

$$Rf = a(t)f(t),$$

where $a(t)\in VDF\bigcap VH(\mu)$.

Theorem 3. 2. 6. *If $a(t), b(t), g(t)\in VDF\bigcap VH(\mu)$, a^{-1} exists and $q\|a^{-1}\|_\mu\|b\|_\mu<1$, then the singular integral equation*

$$a(t)f(t)+\frac{b(t)}{\pi i}\int_{C_0}f(\tau)(\zeta(\tau-t))+\zeta(t))d\tau=h(t) \tag{3.2.12}$$

has a unique solution.

Proof. Since a^{-1} exists, therefore R^{-1} exists, As $q\|a^{-1}\|_\mu\|b\|_\mu<1$, it follows that $\|R^{-1}K\|<1$, where $\|T\|=\inf\frac{\|Tf\|_\mu}{\|f\|_\mu}$. Therefore $(J-R^{-1}K)^{-1}$ exists and $V^{-1}=(J-R^{-1}K)^{-1}R^{-1}$, where $V=R-K$ and J is the identity operator. Q. E. D.

Problem: Suppose that C_0 is a closed path with an nonempty interior, and that C_0 is in the fundamental parallelogram P_0 (see Sec. 7) and the point $z=0$ is included in the interior of C_0. Let

$$C=\{t+m\omega_1+n\omega_2 : t\in C_0, m, n \text{ are arbitrary integers}\},$$

and E_0 be a commutative Banach algebra with the unit element I. Functions $\Phi(z)$ in VDF are found, piecewise regular and with poles of order at most d at $z=0$ and $z=m\omega_1+n\omega_2$ and which satisfy the boundary condition

$$\Phi^+(t)=\Phi^-(t)G(t)+g(t) \text{ on } C_0, \tag{3.2.13}$$

where $G(t)$, $g(t)\in VDF\bigcap VH(\mu)$, with $G(t)\neq\theta$, θ the zero element in E_0. The type of problem (3. 2. 13) is denoted by VDR.

Theorem 3. 2. 7. *If*

$$\frac{1}{2\pi i}\int_{C_0}\Phi^-(t)\zeta(t)dt=0, \tag{3.2.14}$$

$$h^{-1},\ (a+b)^{-1} \text{ exist on } C_0, \tag{3.2.15}$$

then VDR (3. 2. 13) *is equivalent to the singular integral equation*(3. 2. 12).

Proof. If we let

$$G(t)=(a(t)-b(t))(a(t)+b(t))^{-1},$$

$$g(t)=h(t)(a(t)+b(t))^{-1},$$

$$\Phi(z)=\frac{1}{2\pi i}\int_{C_0}(\Phi^+(t)-\Phi^-(t))(\zeta(t-z)+\zeta(z))dt,$$

then we get equivalent relations between (3. 2. 12) and (3. 2. 13). Q. E. D.

Suppose that a^{-1} exists, $q\|b\|_\mu\|b\|a^{-1}\|_\mu<1$, h, a, b, Φ^- satisfy conditions (3. 2. 14), (3. 2. 15) and

$$\Phi(z)=\frac{1}{2\pi i}\int_{C_0}f(\tau)\ (\zeta(t-z)+\zeta(z))dt, \tag{3. 2. 16}$$

where $f(t)$ is the solution of (3. 2. 13). From Theorem 3. 2. 14 we conclude that VDR(3. 2. 13) has the unique solution(3. 2. 16).

Similarly, we can discuss the singular integral equation with parameter λ, given by

$$\lambda\ a(t)f(t)+\frac{b(t)}{\pi i}\int_{C_0}f(\tau)(\zeta(\tau-t)+\zeta(t))d\tau=h(t). \tag{3. 2. 17}$$

Obviously, the equation (3. 2. 17) becomes

$$\lambda Rf-Kf=h.$$

If $\lambda\in\rho(R^{-1}K)$, where $\rho(R^{-1}K)$ is the resolvent set of $R^{-1}K$, then the integral equation (3. 2. 17) has the unique solution

$$f=R(\lambda;R^{-1}K)R^{-1}h.$$

Similarly, we can devive the solution of the following VDR

$$\Phi^+(t,\lambda)=\Phi^-(t,\lambda)G(t,\lambda)+g(t,\lambda) \tag{3. 2. 18}$$

with a parameter λ, where

$$G(t,\lambda)=(\lambda a(t)-b(t))(\lambda a(t)+b(t))^{-1},$$

$$g(t,\lambda)=h(t)(\lambda a(t)+b(t))^{-1}.$$

Here the description is omitted.

2. 4. VECTOR-VALUED BOUNDARY VALUE PROBLEMS WITH THE BOUNDARY A STRAIGHT LINE

Suppose that E_0 is a Banach algebra , that the vector-valved function $\Phi_1(z)$ may have a pole at the point a_0, which is not on the real axis. $\Phi_1(z)$ is bounded and regular except on a neighborhood of the point a_0 and satisfies

$$\Phi_1^+(t)=\Phi_1^-(t)G_1(t)+g_1(t) \qquad \text{on the real axis}, \tag{3.2.19}$$

where $G_1(t), G_1(t), \Phi_1^\pm(t)\in DVH(\mu)$, $G_1(t)\neq\theta$, $G_1^{-1}(t)$ exists on the real axis , and $G_1(z)$ is regular on the upper half plane. Without loss of generality, we will assume $a_0=-i$ and $G_1(z)$ has no zero on the upper half plane.

We first study the homogeneous version of the problem (3.2.19)

$$\Phi_1^+(t)=\Phi_1^-(t)G_1(t) \qquad \text{on the real axis.} \tag{3.2.20}$$

We consider the following formal expression as a particular solution of (3.2.20)

$$\Phi_0(z)=I+\int_{-\infty}^{+\infty}\frac{tz+1}{t-z}\varphi(t)dt, \tag{3.2.21}$$

where $\varphi(t)$ is an unknown function with $\varphi(t)\in VH(\mu)$, and I is the unit element. From (3.2.14) we obtain the boundary values

$$\Phi_0^\pm=I\pm\pi i(t^2+1)\varphi(t)+\int_{-\infty}^{+\infty}\frac{t\tau+1}{\tau-t}\varphi(\tau)d\tau, \tag{3.2.22}$$

where $\Phi_0^+(t)$ and $\Phi_0^-(t)$ are boundary values of $\Phi_0(z)$ as z tends to a point t on the real axis from the upper half plane and the lower half plane respectively.

If $\tilde{z}=w(z)=\dfrac{-zi}{z+i}$, then w maps the real axis onto the circle C. Let us denote points on the circle by t, and let

$$\tilde{\varphi}(\tilde{t})=f\left(-\frac{\tilde{t}i}{\tilde{t}+i}\right) \text{ and } \tilde{\Phi}_0(\tilde{z})=\tilde{\Phi}_0\left(-\frac{\tilde{z}i}{\tilde{z}ti}\right).$$

From (3.2.22) we deduce

$$\Phi_0^\pm\left(\frac{-\tilde{t}i}{\tilde{t}+i}\right)=I\pm\pi i\tilde{\varphi}(\tilde{t})\frac{2\tilde{t}i-1}{(\tilde{t}+i)^2}+\int_C\frac{\tilde{t}i+\tilde{\tau}i-1}{\tilde{\tau}-\tilde{t}}\frac{\tilde{\varphi}(\tilde{t})}{(\tilde{\tau}+i)^2}d\tilde{\tau}. \tag{3.2.23}$$

Let

$$\begin{gathered}
\tilde{G}_1(\tilde{t})=G_1\left(-\frac{\tilde{t}i}{\tilde{t}+i}\right),\\
\tilde{\Phi}_0^\pm(\tilde{t})=\Phi_0^\pm\left(-\frac{\tilde{t}i}{\tilde{t}+i}\right),\\
\tilde{a}(\tilde{t})=\pi i\frac{2\tilde{t}i-1}{(\tilde{t}+i)^2}(I+\tilde{G}(\tilde{t})),\\
K(\tilde{t},\tilde{\tau})=\frac{\tilde{t}i+\tilde{\tau}i-1}{(\tilde{\tau}+i)^2}(I-\tilde{G}(\tilde{t}))\pi i,\\
\tilde{f}(\tilde{t})=\tilde{G}_1(\tilde{t})-I,
\end{gathered} \tag{3.2.24}$$

$$\tilde{b}(t)=K(\tilde{t},\tilde{t}).$$

We replace (3.2.20) by (3.2.24). Thus

$$\tilde{a}(\tilde{t})\tilde{\varphi}(\tilde{t})+\frac{1}{\pi i}\int_C\frac{K(\tilde{t},\tilde{\tau})}{\tilde{\tau}-\tilde{t}}\tilde{\varphi}(\tilde{t})d\tilde{\tau}=\tilde{f}(\tilde{t}) \text{ on } C. \tag{3.2.25}$$

Clearly,

$$\tilde{a}(\tilde{t})+\tilde{b}(\tilde{t})=\tilde{t}-\frac{1}{2i}\tilde{S}(\tilde{t}), \tag{3.2.26}$$

$$\tilde{a}(\tilde{t})-\tilde{b}(\tilde{t})=\tilde{t}-\frac{1}{2i}\tilde{D}(\tilde{t}), \tag{3.2.27}$$

where $\tilde{S}(\tilde{t})=-\frac{4\pi I}{(\tilde{t}+i)^2}$ and $\tilde{D}(\tilde{t})=-\frac{4\pi\tilde{G}_1(\tilde{t})}{(\tilde{t}+i)^2}$.

$(\tilde{a}(\tilde{t})\pm\tilde{b}(\tilde{t}))^{-1}$ exists, because $(G_1(t))^{-1}$ exists. $\tilde{G}_1(\tilde{z})$ is regular inside C, since $G_1(z)$ is regular in the upper half plane. The only zero of $\tilde{a}(\tilde{z})-\tilde{b}(\tilde{z})$ is $\beta=\frac{1}{2i}$.

We write down some obvious results for the solution of the integral equation (3.2.25) as follows:

(i) $(\frac{d}{d\tilde{z}}(\tilde{a}(\beta)\pm\tilde{b}(\beta)))^{-1}$ *exists.*

(ii) *If* $\varphi(t)\in VH(\mu)$, *then*

$$\tilde{\psi}(\tilde{t})=\frac{1}{2\pi i}\int_C\frac{K(\tilde{t},\tilde{\tau})}{\tilde{\tau}-\tilde{t}}\tilde{\varphi}(\tilde{\tau})d\tilde{\tau}\in VH(\mu).$$

(iii) *If* $\varphi(t)\in VH(\mu)$, *then*

$$\tilde{\psi}(\tilde{z})=\frac{1}{2\pi i}\int_C\frac{K(\tilde{z},\tilde{t})}{\tilde{\tau}-\tilde{t}}\tilde{\varphi}(\tilde{\tau})d\tilde{\tau} \tag{3.2.28}$$

is regular inside C and)

$$\tilde{\psi}^+(\tilde{t})=\frac{1}{2}\pi i\tilde{\varphi}(\tilde{t})\frac{(2\tilde{t}i-1)(I-G_1(\tilde{t}))}{(\tilde{t}+i)^2}+\frac{1}{2\pi i}\int_C\frac{K(\tilde{t},\tilde{\tau})}{\tilde{\tau}-\tilde{t}}\tilde{\varphi}(\tilde{\tau})d\tilde{\tau}. \tag{3.2.29}$$

Form (3.2.25) and (3.2.28) we get

$$\tilde{\varphi}(\tilde{t})=(\tilde{f}(\tilde{t})-2\tilde{\varphi}^+(\tilde{t}))(\tilde{a}(\tilde{t})-\tilde{b}(\tilde{t}))^{-1} \text{ for } \tilde{t}\in C. \tag{3.2.30}$$

Let

$$\tilde{B}_0(\tilde{t})=\frac{K(\tilde{t},\beta)\tilde{D}^{-1}(\beta)}{\beta-\tilde{\tau}},$$

$$b'(\beta)=K'_{\tilde{\tau}}(\beta,\beta)+k'_{\tilde{t}}(\beta,\beta)=k'_1+k'_2,$$

where $K'_j=4\pi(I-\tilde{G}_1(\beta))$ *for* j=1,2.

Here there are two cases:

(i) If $\tilde{G}_1(\beta)=-I$ and $(I-\tilde{G}_1(\beta))^{-1}$ exist, then

$$\tilde{\varphi}(\tilde{t})=\tilde{a}(\tilde{t})\tilde{f}(\tilde{t})(\tilde{a}^2(\tilde{t})-\tilde{b}^2(\tilde{t}))^{-1}$$

$$-\frac{(\tilde{a}(\tilde{t})+\tilde{b}(\tilde{t}))^{-1}}{\pi i}\int_C\frac{K(\tilde{t},\tilde{\tau})}{\tilde{\tau}-\tilde{t}}\tilde{f}(\tilde{\tau})(\tilde{a})(\tilde{\tau})-\tilde{b}(\tilde{\tau})^{-1}d\tilde{\tau} \tag{3.2.31}$$

$$+\tilde{C}\,\tilde{B}_0(\tilde{t})(\tilde{a}(\tilde{t})+\tilde{b}(\tilde{t}))^{-1},$$

where $\tilde{C}=\frac{2\tilde{D}(\beta)}{\pi i}(\tilde{S}(\beta)-2ki')^{-1}\int_C\frac{k(\beta,\tilde{\tau})\tilde{f}(\tilde{\tau})(\tilde{a}(\tilde{\tau}))-\tilde{b}(\tilde{\tau}))^{-1}}{\tilde{\tau}-\beta}d\tilde{\tau}.$

Proof. we replace (3.2.28) by (3.2.30). Then

$$\tilde{\psi}(\tilde{z})=\frac{1}{2\pi i}\int_C\frac{K(\tilde{z},\tilde{\tau})\tilde{f}(\tilde{\tau})(\tilde{a}(\tilde{\tau}-\tilde{b}(\tilde{\tau}))^{-1}}{\tilde{\tau}-\tilde{z}}d\tilde{\tau}$$

$$-\frac{1}{\pi i}\int_C\frac{K(\tilde{z},\tau)(\tilde{a}(\tilde{\tau})-\tilde{b}(\tilde{\tau}))^{-1}}{\tilde{\tau}-\tilde{z}}d\tilde{\tau}$$

$$=\frac{1}{2\pi i}\int_C\frac{K(\tilde{z},\tilde{\tau})\tilde{f}(\tilde{\tau})(\tilde{a}(\tilde{\tau})-\tilde{b}(\tilde{\tau}))^{-1}}{\tilde{\tau}-\tilde{z}}d\tilde{\tau}$$

$$-2\tilde{b}(\tilde{z})\tilde{\psi}(\tilde{z})(\tilde{a}(\tilde{z})-\tilde{b}(\tilde{\tau}))^{-1})_{\tilde{\tau}=\beta}$$

$$-2\,res\left(\frac{K(\tilde{z},\tilde{\tau})\tilde{\psi}^+(\tilde{\tau})(\tilde{a}(\tilde{\tau})-\tilde{b}(\tilde{\tau}))^{-1}}{\tilde{\tau}-\tilde{z}}\right)_{\tilde{\tau}=\beta}.$$

We let

$$\tilde{F}(\tilde{z})=\frac{1}{2\pi i}\int_C\frac{K(\tilde{z},\tilde{\tau})\tilde{f}(\tilde{\tau})(\tilde{a}(\tilde{\tau})-\tilde{b}(\tilde{\tau}))}{\tilde{\tau}-\tilde{z}}d\tilde{\tau},$$

then

$$(\tilde{a}(\tilde{z})+\tilde{b}(\tilde{z}))\tilde{a}(\tilde{z})-\tilde{b}(\tilde{z})))^{-1}\tilde{\psi}(\tilde{z})$$

$$=\tilde{F}(\tilde{z})-2\,res\,(\frac{K(\tilde{z},\tilde{t})(\tilde{a}(\tilde{t})-\tilde{b}(\tilde{t})}{\tilde{t}-\tilde{z}})_{\tilde{\tau}=\beta}\tilde{\psi}(\beta).$$

Hence

$$\tilde{\psi}(\tilde{z})=(\tilde{a}(\tilde{z})-\tilde{b}(\tilde{z}))(\tilde{a}(\tilde{z}+\tilde{b}(\tilde{z}))^{-1}(\tilde{F}(\tilde{z})-2\tilde{B}_0(\tilde{z})\tilde{\psi}(\beta)), \tag{3.2.32}$$

where

$$\tilde{B}_0(\tilde{z})=res(\frac{K(\tilde{Z},\tilde{T})(\tilde{a}(\tilde{\tau})-\tilde{b}(\tilde{\tau}))^{-1}}{\tilde{t}-\tilde{z}})_{\tilde{\tau}=\beta}$$

$$=(\frac{K(\tilde{z},\tilde{\tau})\tilde{D}(\tilde{\tau})}{\tilde{\tau}-\tilde{z}})_{\tilde{\tau}=\beta}.$$

Obviously both of the following two conditions should be satisfied in order that the equation (3.2.25) has a solution:

Condition (i)(3. 2. 32) holds as $z=\beta$

Condition (ii) the right side of (3. 2. 32) is regular in the interior of C.

Under condition(1), we let $\tilde{c}=\tilde{\psi}(\beta)$ be the constant element. From the equation (3. 2. 32) with $\tilde{z}=\beta$, we can get $\tilde{c}$, then let z in the interior tend to t and replace $\tilde{z}$ by t in (3. 2. 32). Therefore we obtain the representation of $\tilde{\psi}^+(\tilde{t})$ by substituting this into (3. 2. 30) we derive(3. 2. 31). Q. E. D.

Under condition (ii) if $\tilde{G}_1(\beta)=-I$, we have the following results:

(1) *If*

$$\int_C \frac{K(\beta,\tilde{\tau})\tilde{f}(\tilde{\tau})(\tilde{a}(\tilde{\tau})-\tilde{b}(\tilde{\tau}))^{-1}}{\tilde{\tau}-\beta}d\tilde{\tau}=0,$$

then the general solution of the equation (3. 2. 25) *is* (3. 2. 31), *where* $\tilde{c}$ *is any constant element*

(2) *If*

$$\int_C \frac{K(\beta,\tilde{\tau})\tilde{f}(\tilde{\tau})(\tilde{a}(\tilde{\tau})-\tilde{b}(\tilde{\tau}))^{-1}}{\tilde{\tau}-\beta}d\tilde{\tau}\neq 0,$$

then the equation (3. 2. 25) *has no solution.*

From $\tilde{f}(\tilde{t})$ we get $\varphi(t)$, hence $\Phi_0(z)$ is derived.

Theorem 3. 2. 8. *If* $(\Phi_0(z))^{-1}$ *and* $(\Phi_0^{\pm}(t))^{-1}$ *exist for* $-\infty<t<+\infty$, *then the general solution of the problem* (3. 2. 20) *is* $\Phi_1(z)=\Phi_0(z)P(z)$, *where* $P(z)$ *is any vector-valued polynomial in* $-\frac{iz}{z+i}$.

Proof. The Möbius transformation $\tilde{t}=-\frac{it}{t+i}$ maps the real axis onto the circle C, and problem(3. 2. 20) becomes:

$$\tilde{\Phi}_1^+(\tilde{t})=\tilde{\Phi}_1^-(\tilde{t})\tilde{G}_1(\tilde{t}) \text{ on } C. \tag{3. 2. 33}$$

Suppose $\tilde{\Phi}_1(\tilde{z})$ is any solution of the problem(3. 11. 15). We have

$$\tilde{\Phi}_1^+(\tilde{t})(\tilde{\Phi}^+(\tilde{t}))^{-1}=\tilde{\Phi}_1^-(\tilde{t})(\tilde{\Phi}_0^-(\tilde{t}))^{-1} \quad \text{on } C.$$

$\tilde{\Phi}_1(\tilde{z})(\tilde{\Phi}_0(\tilde{z}))^{-1}$ is a vector-valued regular function on the complex plane by the vector-valued Cauchy Theorem. Because $\tilde{\Phi}_1(\tilde{z})(\tilde{\Phi}_0(\tilde{z}))^{-1}$ may have a pole at $\tilde{z}=\infty$, $\tilde{\Phi}_1(\tilde{z})(\tilde{\Phi}_0(\tilde{z}))^{-1}=\tilde{P}(\tilde{z})$, where $\tilde{P}(\tilde{z})$ is any vector-valued polynomial in $\tilde{z}$. Q. E. D.

Theorem 3. 2. 9. *Under the same hypotheses as in Theorem* 3. 2. 8, *the general solution of* (3. 2. 19) *is*

$$\Phi_1(z)=\frac{z+i}{2\pi i}\Phi_0(z)\int_{-\infty}^{+\infty}\frac{g_1(\tau)(\Phi_0^+(\tau))^{-1}}{(\tau-z)(\tau+i)}d\tau+P(z)\Phi_0(z),$$

where $P(z)$ *is any vector-valued polynomial in* $-\frac{zi}{z+i}$.

Proof. The problem(3. 2. 19) becomes

$$\tilde{\Phi}_1^+(\tilde{t})=\tilde{\Phi}^-(\tilde{t})\tilde{G}_1(\tilde{t})+\tilde{g}_1(\tilde{t}) \text{ on } C \tag{3. 2. 34}$$

after application of the Möbius transformation $w=-\frac{it}{t+i}$, where $\tilde{g}_1=g_1(-\frac{it}{t+i})$. From (3. 2. 34)we get

$$\tilde{\Phi}^+(\tilde{t})(\tilde{\Phi}_0^+(\tilde{t}))^{-1}=\tilde{\Phi}_1^-(\tilde{t})(\tilde{\Phi}_0^-(\tilde{t}))^{-1}+\tilde{g}_1(\tilde{t})(\tilde{\Phi}_0^+(\tilde{t}))^{-1}. \tag{3. 2. 35}$$

Hence $\tilde{\Phi}_1(\tilde{z})(\tilde{\Phi}_0(\tilde{z}))^{-1}$ is piecewise regular in the whole complex plane except possibly for a pole at $\tilde{z}=\infty$.

From the vector-valued Plemelj formula the solution of the problem (3. 2. 35) can be expressed as

$$\tilde{\Phi}_1(\tilde{z})(\tilde{\Phi}_0(z))^{-1}=\frac{1}{2\pi i}\int_C\frac{\tilde{g}_1(\tilde{\tau})(\tilde{\Phi}_0^+(\tilde{\tau}))^{-1}}{\tilde{\tau}-\tilde{z}}d\tilde{\tau}+\tilde{P}(\tilde{z}),$$

where $\tilde{P}(\tilde{z})$ is a vector-valued polynomial in $\tilde{z}$

We get the required result for E, t from $\tilde{z}$, $\tilde{t}$ recpectively. Q. E. D.

If there is a mapping $t_1=u(t)$,which maps the upper half-plane into the interior of C_0 in Sec. 10,and if its inverse mapping is an elliptic function $t=v(t_1)$, then *VDR* (3. 2. 13) becomes

$$\Phi^+(u(t))=\Phi^-(u(t))G(u(t))+g(u(t)) \text{ on the real axis.}$$

Theorem 3. 2. 10. *If* $G(z)$ *is regular in the interior of* C_0 *and if* $(\Phi_0(z))^{-1}$*and* $(\Phi^\pm(t))^{-1}$*exist, then any solutin of VDR* (3. 2. 13) *can be expressed as*:

$$\Phi(\zeta)=\frac{v(\zeta)+i}{2\pi i}\Phi_0(v(\zeta))\int_{C_0}\frac{g(\tau_1)(\Phi_0^+(v(\tau_1)))^{-1}v'(\tau_1)}{(v(\tau_1)-v(\zeta))(v(\tau_1)-i)}d\tau_1+P(v(\zeta))\Phi_0(v(\zeta)).$$

Proof. We obtain the required result by substituting ζ,τ_1 for z,τ in Theorem 3. 2. 9 respectively, where $z=v(\zeta)$, $\tau=v(\tau_1)$, $\Phi(\zeta)=\Phi_1(v(\zeta))$ *and* $g(\tau_1)=g_1(v(\tau_1))$. Q. E. D.

2.5. THE SOLUTION OF THE VECTOR-VALUED DISTURBANCE PROBLEM

Definition 3.2.2. If $\Phi_n^{\pm}, \Phi^{\pm}, G, g, g_n, \Delta_n \in DVH(\mu)$ for $t \in C_0$, the following vector-valued doubly-periodic Riemann boundary value problem

$$\Phi_n^+(t) - \Phi_n^-(t) = \Phi_n^-(t)(G(t) + \Delta_n(t)) + g_n(t) \text{ on } C_0 \tag{3.2.36}$$

is called a disturbance of

$$\Phi^+(t) - \Phi^-(t) = \Phi^-(t)G(t) + g(t) \text{ on } C_0, \tag{3.2.37}$$

where $DVH(\mu) = VDF \cap VH(\mu)$, $\|g_n - g\|_\mu \to 0$, $\|\Delta_n\|_\mu \to 0$ as $n \to \infty$, and the vector-valued doubly-periodic piecewise regular functions $\Phi_n(z)$ satisfy (3.2.36) on C_0 which have at most simple poles at $z=0$ and $z = m\omega_1 + n\omega_2$.

From the vector-valued Pl'emelj formula and the properties of Weierstrass' s ζ-function we can obtain the following result.

Proposition 3.2.1. *Suppose*

$$\psi(z) = \frac{1}{2\pi i}\int_{C_0} f(t)(\zeta(t-z) + \zeta(z))dt,$$

where $f \in DVH(\mu)$, *then*

$$\psi^{\pm}(t) = \pm\frac{1}{2}f(t) + \frac{1}{2\pi i}\int_{C_0} f(t)(\zeta(t-t) + \zeta(\tau))dt. \tag{3.2.38}$$

For given $G, \Delta_n, g_n \in DVH(\mu)$ in (3.2.36), let

$$\Phi_{n,k}(z) = \frac{1}{2\pi i}\int_{C_0} ((G(t) + \Delta_n(t))\Phi_{n,k-1}^-(t) + g_n(t))(\zeta(t-z) + \zeta(z))dt \tag{3.2.39}$$

for $k = 0, 1, 2, \cdots$.

Set $\Phi_{n,0}^-(t) = \theta$. We obtain

$$\Phi_{n,k+1}(z) - \Phi_{n,k}(z)$$
$$= \frac{1}{2\pi i}\int_{C_0} (G(t) + \Delta_n(t))(\Phi_{n,k}^-(t) - \Phi_{n,k-1}^-(t))(\zeta(t-z) + \zeta(t))dt.$$

From Theorem 3.2.5 and Proposition 3.2.1, we obtain

$$\|\Phi_{n,k+1}^- - \Phi_{n,k}^-\|_\mu \leqslant q\|G + \Delta_n\|_\mu \|\Phi_{n,k}^- - \Phi_{n,k-1}^-\|_\mu$$
$$\leqslant (q\|G + \Delta_n\|_\mu)^k \|\Phi_{n,1}^-\|_\mu. \tag{3.2.40}$$

From (3.2.39) we get

$$\Phi_{n,1}^-(t) = -\frac{1}{2}g_n(t) + \frac{1}{2\pi i}\int_{C_0} g_n(t)(\zeta(\tau - t) + \zeta(t))dt.$$

Since $\{g_n\}$ is bounded in $DVH(\mu)$,

$$\|\Phi_{n,1}^-\|_\mu \leq m_2, \tag{3.2.41}$$

where m_2 is a constant. If

$$q\|G+\Delta_n\|_\mu < 1 \text{ for } n>N, \tag{3.2.42}$$

by (3.2.40) $\lim_{k\to\infty} \Phi_{n,k}^-(t)$ exists for any $t\in C_0$ and $n=1,2,\cdots$, and is denoted by $\Phi_n^-(t)$, thus

$$\|\Phi_{n,k}^- - \Phi_n^-\|_\mu \to 0 \text{ as } k\to\infty. \tag{3.2.43}$$

Similarly,

$$\|\Phi_{n,k}^+ - \Phi_n^+\|_\mu \to 0 \text{ as } k\to\infty. \tag{3.2.44}$$

Equation (3.2.39) and the vector-valued Plemelj formula give

$$\Phi_n^\pm(t) = \pm\frac{1}{2}((G(t)+\Delta_n(t))\Phi_n^-(t)+g_n(t))$$

$$+\frac{1}{2\pi i}\int_{C_0}((G(t)+\Delta_n(t))\Phi_n^-(\tau)+g_n(t))(\zeta(\tau-t)+\zeta(t))d\tau. \tag{3.2.45}$$

Therefore

$$\Phi_n(z) = \frac{1}{2\pi i}\int_{C_0}((G(t)+\Delta_n(t))\Phi_n^-(t)+g_n(t))(\zeta(t-z)+\zeta(z))dt$$

satisfies the boundary value problem (3.2.36). We therefore obtain the following conclusion.

Proposition 3.2.2. *If* $q\|G+\Delta_n\|_\mu<1$, *then the disturbance problem* (3.2.36) *of the vector-valued doublyperiodic Riemann boundary value problem* (3.2.37) *with at most a pole at* $z=0$ *of order* 1 *has a unique solution, which is obtained by the above step-by-step method.*

Now we are ready find the relation between the solution of the vector-valued singular integral equation

$$a_n(t)y_n(t)+\frac{b_n(t)}{\pi i}\int_{C_0} y_n(\tau)(\zeta(\tau-t)+\zeta(t))d\tau = f_n(t) \tag{3.2.46}$$

and the solution of the disturbance problem (3.2.36), where $a_n, b_n, y_n, f_n \in DVH(\mu)$ for $n=1,2,\cdots$.

Proposition 3.2.3. *If*

$$(a_n+b_n)^{-1} \tag{3.2.47}$$

exists, then the solution of the integral equation (3. 2. 46) *and the solution of the disturbance problem* (3. 2. 36) *with the pole at* $z=0$ *of order at most* 1 *are equivalent under the condition*

$$\frac{1}{2\pi i}\int_{C_0}\Phi_n^-(t)\zeta(t)dt=0.$$

Proof. Suppose

$$g_n=f_n(a_n+b_n)^{-1},$$
$$\Delta_n=-G-2b_n(a_n+b_n)^{-1}. \tag{3.2.48}$$

If $y_n(t)$ is the solution of integral equation (3. 2. 46), set

$$\Phi_n(z)=\frac{1}{2\pi i}\int_{C_0}y_n(t)(\zeta(t-z)+\zeta(z))dt. \tag{3.2.49}$$

From (3. 2. 49) and Proposition 3. 2. 1 we get

$$\Phi_n^+(t)-\Phi_n^-(t)=y_n(t),$$
$$\Phi_n^+(t)+\Phi_n^-(t)=\frac{1}{\pi i}\int_{C_0}y_n(\tau)(\zeta(\tau-t)+\zeta(t))dt. \tag{3.2.50}$$

The integral equation (3. 2. 46) becomes

$$\Phi_n^+(t)-\Phi_n^-(t)=\Phi_n^-(t)(G(t)+\Delta_n(t))+g_n(t).$$

The remainder of the proof is easy, by using the basic properties of the ζ-function.

Q. E. D.

Suppose operators A_n, A are defined by the formula

$$A_ny\equiv a_n(t)y(t)+\frac{b_n(t)}{\pi i}\int_{C_0}y(\tau)(\zeta(\tau-t)+\zeta(t))d\tau,$$
$$Ay\equiv a(t)y(t)+\frac{b(t)}{\pi i}\int_{C_0}y(\tau)(\zeta(\tau-t)+\zeta(t))d\tau,$$

where $a_n, a, b_n, b, y\in DVH(\mu)$.

Obviously A_n, A are bounded linear operators on the space $DVH(\mu)$.

For an arbitrarily small positive number r, set

$$H_r=\{(a_n,b_n):\|A_n\|\geq r\}.$$

Proposition 3. 2. 4. *If*

$$\|a_n-a\|_\mu\to 0,\ \|b_n-b\|_\mu\to 0,\ \|f_n-f\|_\mu\to 0 \text{ as } n\to\infty \tag{3.2.51}$$

and there exist r, N *such that*

$$(a_n,b_n)\in H_r \text{ for } n>N, \tag{3.2.52}$$

then the following results hold:

(i) *The integral equation* (3. 2. 46) *has a unique solution* y_n.

(ii) *The integral equation*

$$a(t)y(t)+\frac{b(t)}{\pi i}\int_{C_0} y(\tau)(\zeta(\tau-t)+\zeta(t))d\tau=f(t)$$

has a unique solution y.

(iii) $\|y_n-y\|_\mu\to 0$ as $n\to\infty$.

Proof. Since (i) and (ii) are obvious by Theorem 2. 11. 6 in [11], we only prove (iii). We have that

$$\|(A_n-A)y\|$$

$$=\|(a_n(t)-a(t))y(t)+\frac{b_n(t)-b(t)}{\pi i}\int_{C_0} y(\tau)(\zeta(\tau-t)-\zeta(t))dt\|$$

$$\leqslant\|a_n-a\|_\mu\|y\|_\mu+c\|b_n-b\|_\mu\|y\|_\mu,$$

$$\|A_n-A\|\leqslant\|a_n-a\|_\mu+c\|b_n-b\|_\mu \tag{3. 2. 53}$$

where c is a constant.

On the other hand,

$$A_n^{-1}-A^{-1}=A_n^{-1}(A-A_n)A^{-1},$$

and $\|A_n^{-1}\|$ is bounded. By (3. 2. 51) and (3. 2. 53) we have

$$\|A_n^{-1}-A^{-1}\|\to 0 \text{ as } n\to\infty,$$

i. e.

$$\|y_n-y\|_\mu\to 0 \text{ as } n\to\infty.$$

Proposition 3. 2. 5. *If conditions* (3. 2. 11), (3. 2. 52), (3. 2. 50) *and* (3. 2. 53) *hold and*

$$\|2b_n(a+b_n)^{-1}+G\|_\mu\to 0 \text{ as } n\to\infty, \tag{3. 2. 54}$$

then

$$\lim_{n\to\infty}\Phi_n(z)\equiv\Phi(z)=\lim_{n\to\infty}\frac{1}{\pi i}\int_{C_0} y_n(\tau)(\zeta(\tau-z)+\zeta(z))dt$$

is the solution of the boundary value problem (3. 2. 55) *and*

$$\Phi(z)=\frac{1}{\pi i}\int_{C_0} y(\tau)(\zeta(\tau-z)+\zeta(z))d\tau. \tag{3. 2. 55}$$

Proof. From Proposition 3. 2. 4, we obtain that the integral equation (3. 2. 46) has unique solution. By Proposition 3. 2. 3, the disturbance problem (3. 2. 36) has a unique solution $\Phi_n(z)$. From Proposition 3. 2. 4 and (3. 2. 49) we conclude that (3. 2. 55) holds and satisfies the boundary value problem (3. 2. 40). Q. E. D.

2. 6. THE VECTOR-VALUED BOUNDARY VALUE PROBLEM IN l_p

In 1979, *J. K. Lu solved the doubly-periodic Riemann boundary value problem with the domain and the range in a complex plane* $\mathscr{C}$ (see [25]). In 1988, C. G. Hu extended above problem to vector valued cases with the domain in $\mathscr{C}$ and the range in a Banach space and got some unique solution of this problem under stronger conditions (see [19]).

In this section, under weaker conditions we obtain general solutions (of above problem) belonging to a Banach space U.

Let $S=\{w: |w|<1\}$ and $\bar{S}=\{w: |w|\leqslant 1\}$. *Set*

$$U=\{f: f \text{ is regular in } S \text{ and continuous on } \bar{S}\}$$

with the norm

$$\|f\|_U = sup\{|f(w)|: w\in \bar{S}\}$$

and the multiplication $f * g = f\cdot g$. Obviously, U is a complete Banach algebra.

Suppose that VDF is the set of all vector-valued doubly periodic continuous functions with periods ω_1 and ω_2, and ranges are elements in U, that $DVH(\mu)$ is the set of all vector valued doubly-periodic functions which satisfy the Hölder condition, that l_0 is a closed path and has a non-empty interior, that the track l_0^* of l_0 is in the fundamental parallelogram

$$P_0=\{\frac{1}{2}(2t_1-1)\omega_1+\frac{1}{2}(2t_2-1)\omega_2: 0\leqslant t_j\leqslant 1 \text{ for } j=1,2\}$$

and the point $z=0$ is included in the interior of l_0, that

$$l=\{t+m\omega_1+n\omega_2: t\in l_0^*,\ m \text{ and } n \text{ are arbitrary integers}\},$$

and that the inner domain of l_0 is denoted by S^+ and

$$S^-=P_0-(S^+\cup l_0^*).$$

Functions $\Phi(z)$ in VDF are found to be piecewise regular and with poles of order at most d at $z=0$ and $z=m\omega_1+n\omega_2$ and which satisfy the boundary condition

$$\Phi^+(t)=\Phi^-(t) * G(t)+g(t) \text{ as } t\in l, \tag{3. 2. 56}$$

where $\Phi^{\pm}(t), g(t)\in DVH(\mu)$ and $G(t)\in DVH(1)$.

Let

$$G(t,w)=\sum_{n=0}^{\infty}a_n(t)w^n \quad \text{as } (t,w)\in l_0^*\times S, \tag{3.2.57}$$

and

$$\|a\|_{H(\mu)}=sup\{\,|a(t)|\,;\ t\in l_0^*\,\}$$

$$+sup\{\frac{|a(t_1)-a(t_2)|}{|t_1-t_2|^{\mu}};\ (t_1,t_2)\in l_0^*\times l_0^* \text{ and } t_1\neq t_2\}.$$

Lemma 3.2.2. *Suppose that*

$$\sum_{n=0}^{\infty}\|a_n\|_{H(1)}<\infty, \tag{3.2.58}$$

and that

$$|G(t,w)|<\bar{M}\,(\bar{M} \text{ is a constant}) \text{ as } (t,w)\in l_0^*\times S. \tag{3.2.59}$$

Then the following results hold:

(i) $G(t,w)$ (*i. e.* $G_w(t)$) $\in H(1)$ *for each fixed* $w\in S$, *that is* $G_w(t)$ *satisfying the Lipschitz condition.*

(ii) $G(t,w)$ (*i. e.* $G_t(w)$) $\in U$ *for each fixed* $t\in l_0^*$.

(iii) *If*

$$Re(G(t,w))\neq 0 \text{ as } (t,w)\in l_0^*\times\bar{S}, \tag{3.2.60}$$

then (a)

$$inf\{|G(t,w)|:(t,w)\in l_0^*\times\bar{S}\}>0 \tag{3.2.61}$$

(b) *Let* $k(w)=\frac{1}{2\pi}(arg\ G(t,w))_{l_0}$ (*see* [28]), *then* $k(w)=k$, *a constant.*

(iv) $f(w)=\int_{l_0}G(t,w)dt\in U$.

(v) *Let* $G_*(t,w)=G(t,w)\mu^{-k}(t)$, *then*

$$L(z,w)=\frac{1}{2\pi i}\int_{l_0}\log G_*(t,w)\ \zeta(t-z)dt\in U$$

for each fixed z *as* $z\bar{\in}l_0^*$, *where*

$$\mu(t)=\sigma(t)\sigma(t-\frac{1}{2}\omega_1-\frac{1}{2}\omega_2)/\sigma(t-\frac{1}{2}\omega_1)\sigma(t-\frac{1}{2}\omega_2),$$

$\sigma(t)$ *and* $\zeta(t)$ *are Weierstrass* σ *and* ζ *functions respectively.*

(vi) $L^+(t,w)\in U$ *for each fixed* $t\in l_0^*$.

Proof. (i) Since $G_w(t)=\sum_{n=0}^{\infty}\in a_n(t)w^n$,

$$|G_w(t_1)-G_w(t_2)|\leqslant\sum_{n=0}^{\infty}|a_n(t_1)-a_n(t_2)|\,|w^n|$$

$$\leqslant\sum_{n=0}^{\infty}\|a_n\|_{H(1)}|t_1-t_2|. \tag{3.2.62}$$

(ii) $G_t(w)\in U$, because

$$|G_t(w)|\leqslant\sum_{n=0}^{\infty}|a_n(t)|\leqslant\sum_{n=0}^{\infty}\|a_n\|_{H(t)}. \tag{3.2.63}$$

(iii)(a) If (3.2.61) does not hold, then there exists a double sequence $\{(t_n,w_m)\}\subset l_0^*\times\bar{S}$ $(n,m=1,2,\cdots)$ such that $t_n\to t_0$ and $w_m\to w_0$ as n and $m\to\infty$ respectively and $\lim\limits_{n\to\infty}\lim\limits_{m\to\infty}G_{t_n}(w_m)=0$. From (3.2.62) we obtain

$$\|G_{t_n}-G_{t_0}\|_U\to 0 \text{ as } n\to\infty \tag{3.2.64}$$

and $\{G_t(w)\}$ is normal on $\bar{S}$, hence $\{G_t(w)\}$ is equicontinuous on $\bar{S}$. Therefore for any $\varepsilon>0$, there exist N and M such that

$$|G_{t_n}(w_0)-G_{t_0}(w_0)|<\frac{1}{2}\varepsilon \text{ as } \nu>N,$$

$$|G_{t_n}(w_m)-G_{t_n}(w_0)|<\frac{1}{2}\varepsilon \text{ as } m>M.$$

Thus

$$|G_{t_n}(w_m)-G_{t_0}(w_0)|$$
$$\leqslant|G_{t_n}(w_m)-G_{t_n}(w_0)|+|G_{t_n}(w_0)-G_{t_0}(w_0)|$$
$$\leqslant\frac{1}{2}\varepsilon+\frac{1}{2}\varepsilon=\varepsilon$$

as $n>N$ and $m>M$. Therefore $G(t_0,w_0)=0$. This is in contradicting with (3.2.60). Hence (3.2.61) holds.

(b) From (3.2.62) we show that $G_w(t)$ is a rectifiable curve for each fixed w in $\bar{S}$.

Now we can show $k(w)$ is bounded in $\bar{S}$. In fact, if $k(w)$ isn't bounded, then without loss of generality, we may assume that there exists a sequence $\{w_n\}$ such that

$$\frac{1}{2\pi}(\arg G(t,w_n))_{l_0}\to\infty \text{ and } w_n\to w_0 \text{ as } n\to\infty.$$

Since $\{G_t(w)\}$ is equicontinuous,

$$\|G_{w_n}-G_{w_0}\|\to 0 \text{ as } n\to\infty, \tag{3.2.65}$$

where $\|G_w\|=sup\{|G_w(t)|:t\in l_0^*\}$. From the Lemma 3. 2. 2 (iii) (a) and (3. 2. 65) we can conclude

$$\frac{1}{2\pi}(arg\ G(t,w_n))_{l_0}\to\frac{1}{2\pi}(arg\ G(t,w_0))_{l_0} \text{ as } n\to\infty, \tag{3. 2. 66}$$

hence

$$\frac{1}{2\pi}(arg\ G(t,w_0))_{l_0}=\infty.$$

It follows that there exists a $t_0\in l_0^*$ such that $G(t_0,w_0)=0$. This yields a contradiction, hence $k(w)$ must be bounded.

If there exist two points $w_0^{(1)}$ and $w_0^{(2)}\in\bar{S}$ such that

$$\frac{1}{2\pi}(arg\ G(t,w_0^{(1)}))_{l_0}\neq\frac{1}{2\pi}(arg\ G(t,w_0^{(2)}))_{l_0}.$$

Let $w_1=\frac{1}{2}(w_0^{(1)}+w_0^{(2)})$, without loss of generality, we can assume

$$\frac{1}{2\pi}(arg\ G(t,w_0^{(1)}))_{l_0}\neq\frac{1}{2\pi}(arg\ G(t,w_1))_{l_0}.$$

Similarly let $w_2=\frac{1}{2}(w_0^{(1)}+w_1)$, without loss of generality, we can assume

$$\frac{1}{2\pi}(arg\ G(t,w_0^{(1)}))_{l_0}\neq\frac{1}{2\pi}(arg\ G(t,w_2))_{l_0},$$

$$\cdots \qquad \cdots$$

$$\frac{1}{2\pi}(arg\ G(t,w_0^{(1)}))_{l_0}\neq\frac{1}{2\pi}(arg\ G(t,w_n))_{l_0},$$

where $w_n=\frac{1}{2}(w_0^{(1)}+w_{n-1})$. Thus there exists an integer N such that

$$\frac{1}{2\pi}(arg\ G(t,w_n))_{l_0}=\frac{1}{2\pi}(arg\ G(t,w_0^{(1)}))_{l_0} \text{ as } n>N,$$

because $\frac{1}{2\pi}(arg\ G(t,w_n))_{l_0}$ are integers. This yields a contradiction, hence $k(w)\equiv k$, a constant.

(iv) Obviously $f(w)\in U$.

(v) From Lemma (iv) we have $L(z,w)\in U$ for each fixed $z\in P_0-l_0^*$ by the condition (3. 2. 59).

(vi) From the Plemelj's formula in [28] we can obtain

$$L^+(t,w)=\frac{1}{2}log\ G_*(t,w)+\frac{1}{2\pi i}\int_{l_0}log\ G_*(\tau,w)\zeta(\tau-t)d\tau$$

$$=\frac{1}{2}\log G_*(t,w)+\frac{1}{2\pi i}\int_{l_0}(\log G_*(\tau,w))\sum_{m,n}{}'\left(\frac{1}{\tau-t-w_{m,n}}+\frac{1}{w_{m,n}}+\frac{\tau-t}{w_{m,n}^2}\right)d\tau$$

$$+\frac{1}{2\pi i}\int_{l_0}\frac{\log G_*(\tau,w)-\log G_*(t,w)}{\tau-t}dt+\frac{1}{2\pi i}\log G_*(t,w)\int_{l_0}\frac{d\tau}{\tau-t}, \quad (3.2.67)$$

where $w_{m,n}=m\omega_1+n\omega_2$, m and n are integers, $\sum'$ is denoted the summation except the term of $m=n=0$. From the Lemma 3. 2. 2 (iv) we obtain $L^+(t,w)\in U$, because each term in (3. 2. 66) belongs to U.

Corollary (i). *Suppose that*

$$G_*(w)=\frac{1}{2\pi i}\int_{l_0}\log G_*(t,w)dt,$$

that

$$X(z,w)=\begin{cases}u^k(z)h(z,w)e^{L(z,w)} & as\ z\in S^+,\\ h(z,w)e^{L(z,w)} & as\ z\in S^-\end{cases} \quad (3.2.68)$$

where $h(z,w)=\sigma(z)/\sigma(z-G_*(w))$, *and that*

$$E\cap P_0=\varnothing \quad (3.2.69)$$

where E is the range of $G_*(w)$, *then the following results are valid:*

(a) $X(z,w)\in U$ *for each fixed* $z\in S_0^\pm$.

(b) $X^+(t,w)\in U$ *for each fixed* $t\in l_0^*$.

Corollary (ii). *Suppose that* $g(t,w)=\sum_{n=0}^{\infty}b_n(t)w^n$, *and that*

$$g_t(w)\in U \text{ and } g_w(t)\in DVH(\mu) \text{ as } (t,w)\in l_0^*\times S, \quad (3.2.70)$$

then

$$\int_{l_0}\frac{g(t,w)}{X^+(t,w)}(\zeta(t-z)+\zeta(z))dt\in U \text{ as } z\overline{\in} l_0^*.$$

We use above Lemma 3. 2. 2 and similar method in [25] easily getting the following results.

Theorem 3. 2. 11. *Suppose that* $G(t,w)$ *satisfies conditions in Lemma* 3. 2. 2 *and* (3. 2. 69), *that* $g(t,w)$ *satisfies the condition* (3. 2. 70), *we have the following cases:*

(i) *If* $G_*(w)=m\omega_1+n\omega_2$, *where m and n are integers and* $m^2+n^2\neq 0$.

(a) *Suppose* $k+d>0$, *then a general solution of* (3. 2. 56) *can be expressed as*

$$\Phi(z,w)=\frac{X(z,w)}{2\pi i}\int_{l_0}\frac{g(t,w)}{X^+(t,w)}(\zeta(t-z)+\zeta(z))dt$$
$$+X(z,w)(a_0(w)+a_1(w)\zeta'(z)+\cdots+a_{k+d-1}(w)\zeta^{(k+d-1)}(z)),$$

where $a_j(w)(j=0,1,\cdots,k+d-1)$ *are arbitrary in* U *and* $\Phi(z,w)\in U$ *for each fixed* $z\in S_0^+$.

(b) *Suppose* $k+d=0$, *then the general solutions of* (3.2.56) *under the condition*

$$\frac{1}{2\pi i}\int_{l_0}\frac{g(t,w)}{X^+(t,w)}dt=0 \tag{3.2.71}$$

are in the form:

$$\Phi(z,w)=\frac{X(z,w)}{2\pi i}\int_{l_0}\frac{g(t,w)}{X^+(t,w)}\zeta(t-z)dt+a(w)X(z,w), \tag{3.2.72}$$

where $a(w)$ *is arbitrary in* U.

(c) *Suppose* $k+d=-1$ *and* (3.2.71) *is satisfied, then the solution of* (3.2.56) *is unique and*

$$\Phi(z,w)=\frac{X(z,w)}{2\pi i}\int_{l_0}\frac{g(t,w)}{X^+(t,w)}(\zeta(\tau-z)-\zeta(t))d\tau. \tag{3.2.73}$$

(d) *Suppose that* $k+d<-1$, *and that* (3.2.56) *is satisfied and*

$$\frac{1}{2\pi i}\int_{l_0}\frac{g(t,w)}{X^+(t,w)}\zeta^{(\lambda)}(t)dt=0 \text{ as } \lambda=1,\cdots,-k-d-1,$$

then the solution of (3.2.56) *is* (3.2.73).

(ii) *Suppose that* $G_*(w)\neq m\omega_1+n\omega_2$, *where* m *and* $n=0,\pm1,\pm2,\cdots$, *that* $G_0(w)=G_*(w)(\mathrm{mod}\ \omega_1,\omega_2)$ *and* $G_0(w)\neq0$, *and that the range of* $G_0(w)$ *is in* P_0:

(a) *Suppose* $k+d+1>0$, *then a general solution of* (3.2.56) *can be expressed*

$$\Phi(z,w)=\frac{X(z,w)}{2\pi i}\int_{l_0}\frac{g(t,w)}{X^+(t,w)}(\zeta(t-z)-\zeta(t-G_0(w))+\zeta(z)$$
$$-\zeta(G_0(w)))dt+X(z,w)(a_1(w)(\zeta'(z)-\zeta'(G_0(w)))+\cdots$$
$$+a_{k+d}(w)(\zeta^{(k+d)}(z)-\zeta^{(k+d)}(G_0(w))).$$

where $a_1(w),\cdots,a_{k+d}(w)$ *are arbitrary elements in* U.

(b) *Suppose* $k+d+1=0$ *and* (3.2.71) *is satisfied, then* (3.2.56) *has the unique solution*

$$\Phi(z,w)=\frac{X(z,w)}{2\pi i}\int_{l_0}\frac{g(t,w)}{X^+(t,w)}(\zeta(t-z)-\zeta(t-G_0(w)))dt. \tag{3.2.74}$$

(c) *Suppose* $k+d=-2$ *and*

$$\frac{1}{2\pi i}\int_{l_0}\frac{g(t,w)}{X^+(t,w)}(\zeta(t-G_0(w))-\zeta(t))dt=0 \tag{3.2.75}$$

and (3.2.71) *is satisfied then* (3.2.56) *has the unique solution* (3.2.74).

(d) *Suppose* $k+d<-2$ *and* (3.2.61) *is satisfied and*

$$\frac{1}{2\pi i}\int_{t_0}\frac{g(t,w)}{X^+(t,w)}\zeta^{(j)}(t)dt=0 \quad as\ j=1,\cdots,-k-d-2,$$

then (3.2.56) *has the unique solution* (3.2.74).

Theorem 3.2.12. *If ranges of* $\Phi^{\pm}(t)$, $G(t)$ *and* $g(t)$ *are in* l_1, *besides conditions in Theorem* 3.2.11, *then ranges of solutions of* (3.2.56) *are in* U, *and solutions can be represented by formulas in Theorem* 3.2.11, *under the sense of isometrical isomorphism.*

Proof. Let

$$f(w)=\sum_{n=0}^{\infty}a_n w^n \text{ as } w\in S$$

and

$$V=\{f:\sum_{n=0}^{\infty}|a_n|<\infty\}$$

with the norm $\|f\|_V=\sum_{n=0}^{\infty}|a_n|$.

Obviously, this give a unique one-to-one correspondence between an element $(a_0, a_1,\cdots,a_n,\cdots)$ of l_1 and a function $f(w)$ of V, i. e. V is isometrically isomorphic to l_1. Because $V\subset U$, conclusions of Theorem 3.2.12 hold. Q. E. D.

Suppose a Banach space E with a Schauder basis $\{e_n\}$ (see [23]) which satisfies the following conditions:

(i) If $x\in\sum_{n=1}^{\infty}a_n e_n\in E$, then

$$(x)^{(a)}=\sum_{n=1}^{\infty}a_n^a e_n\in E$$

for any fixed $a\geqslant 1$.

(ii) If $y=\sum_{n=1}^{\infty}b_n e_n\in E$, then

$$(x)(y)=\sum_{n=1}^{\infty}a_n b_n\in E.$$

Clearly, E is a commutative Banach algebra (see [11] Chapter iv).

Theorem 3. 2. 13. *Suppose vector-valued doubly-periodic piecewise regular function* $\psi(z)$, *except for singular point* $z=0$ *and* $z=m\omega_1+n\omega_2$ *at most, satisfies the boundary condition*

$$(\psi^+(t))^{(p)}=(\psi^-(t)\}^{(p)} * G(t)+g(t) \quad \text{for } t\in l_0^*, \tag{3.2.76}$$

where ranges of $\psi^\pm(t)$ *are in* l_0, *ranges of* $G(t)$ *and* $g(t)$ *are in* l_1, *and* $*$ *expresses the convolution in the space* l_1, $G(t)\in DVH(1)$, $g(t)\in DVH(\mu)$, *and they satisfy various corresponding conditions in Theorem* 3. 2. 11, *then the corresponding solutions of* (3. 2. 76) *are*

$$\psi(z)=(\Phi(z))^{(\frac{1}{p})}=\sum_{j=1}^{\infty}\varphi_j^{\frac{1}{p}}(z)e_j$$

and

$$\psi^\pm(t)\in DVH(\mu/p),$$

where $\Phi(z)=\sum_{j=1}^{\infty}\varphi_j(z)e_j$ *is the solution of* (3. 2. 56).

Proof. By Theorem 3. 1. 6 for any j we have

$$\begin{aligned}&|(\varphi_j^\pm(t_1))^{\frac{1}{p}}-(\varphi_j^\pm(t_2))^{\frac{1}{p}}|\\ &\leqslant|\varphi_j^\pm(t_1)-\varphi_j^\pm(t_2)|^{\frac{1}{p}}\\ &\leqslant\|\varphi_j^\pm\|_{H(\mu)}|t_1-t_2|^{\mu/p}.\end{aligned}$$

We use again Theorem 3. 1. 6 to obtain

$$\psi^\pm(t)\in DVH(\mu/p).$$

The proof of the rest of this theorem is clear.

Corollary. *If* $\psi(z)$ *satisfies the conditions of Theorem* 3. 2. 13 *with the ranges of* $\psi^\pm(t)$ *in a Hilbert space, then the boundary value problem* (3. 2. 76) *have various corresponding solution in* U, *under the sense of isometrical isomorphism, where*

$$\psi^\pm(t)\in DVH(\frac{1}{2}\mu).$$

Proof. Let $p=2$ in Theorem 3. 2. 13, we can derive the conclusions as required immediately. Q. E. D.

3. The Analysis of Locally Convex Spaces

3. 1. G-DIFFERENTIABILITY

Definition 3. 3. 1. A set $C^*(x_0)$ is called a C-star about x_0 if $C^*(x_0)=x_0+H$, where $h\in H$, $|z|\leqslant 1$ implies that $zh\in H$.

Definition 3. 3. 2. A subset D of a real (or complex) linear system E is said to be finitely open if for each choice of elements $y_1,\cdots,y_n\in E$, the elements $\sum_1^n a_k y_k$ which are in D correspond to an open subset of the space R_n or Z_n with elements $(a_1,\cdots,a_n)$.

Let $\rho(x,h)=\sup\{\rho;\ |z|\leqslant p$ implies $x+zh\in D\}$, where D is finitely open.

Definition 3. 3. 3. Let $y=f(x)$ be defined in the finitely open set D in values in a Banach space and suppose that for every $x\in D$ and $h\in E$ the quotient

$$\frac{f(x+zh)-f(x)}{z}$$

which is defined for $0<|z|<\rho(x,h)$, tends to a unique limit as $z\to 0$. We then say that

(1) $f(x)$ is G-differentiable in D,

(2) $\delta f(x,h)=\delta_z^h f=\lim\limits_{z\to 0}\dfrac{f(x+zh)-f(x)}{z}$ is the first variation of $f(x)$ with respect to the increment h.

Theorem 3. 3. 1. *If $f(x)$ and $g(x)$ are G-differentiable in D, then so is $\alpha f+\beta g$ and*

$$\delta_z^h(\alpha f+\beta g)=\alpha\delta_z^h f+\beta\delta_z^h g. \tag{3. 3. 1}$$

Proof.

$$\frac{(\alpha f(x+zh)+\beta g(x+zh))-(\alpha f(x)+\beta g(x))}{z}$$

$$=\alpha\frac{f(x+zh)-f(x)}{z}+\beta\frac{g(x+zh)-g(x)}{z}$$

gives (3. 3. 1).

Theorem 3. 3. 2. $\delta(x,h)$ *is homogeneous of degree one in* h, *i. e.*

$$\delta f(x,ah)=a\delta f(x,h). \tag{3. 3. 2}$$

Proof. Let $\eta = az$. We have

$$\frac{f(x+zah)-f(x)}{z} = a\,\frac{f(x+\eta h)-f(x)}{\eta},$$

where $\eta = za$. It follows that (3. 3. 2) holds.

Definition 3. 3. 4. A function $f(x)$ defined on E, with values in F is said to be homogeneous of degree n if

$$f(ax) = a^n f(x).$$

Such a function is called bounded if there exists an $M > 0$ such that

$$\|f(x)\| \leqslant M\|x\|^n, \text{ for all } x \in E.$$

Definition 3. 3. 5. Let $y = f(x)$ on E with values in F be defined in the finitely open set D. We say that $f(x)$ is F-differentiable and possesses a total or Frechet differential in D, if

(i) $\delta f(x,h)$ exists as a bounded homogeneous function of degree one in h;

(ii) $\lim\limits_{\|h\|\to 0} \frac{1}{\|h\|}\|f(x+h)-f(x)-\delta f(x,h)\| = 0$ *for all* $x \in D$.

Theorem 3. 3. 3. *$f(x)$ is G-differentiable in the finitely open set D if and only if for every $x \in D$ and $h \in E$, $f(x+zh)$ is a regular function of z when $|z| < \rho(x,h)$.*

Proof. This follows from the observation

$$\left\{\frac{d}{dz} f(x+(z_0+z)h)\right\}_{z=0} = \lim_{z\to 0}\frac{1}{z}\{f((x+z_0h)+zh)-f(x+z_0h)\}$$

$$= \delta f(x_0+z_0h, h). \qquad \text{Q. E. D.}$$

If $f(x)$ is G-differentiable in the finitely open set D and if $x \in D$ we may define the n-th variation $\delta^n f(x,h)$ of $f(x)$ with increment h as

$$\delta^n f(x,h) = \left[\frac{d^n}{dz^n} f(x+zh)\right]_{z=0}. \tag{3. 3. 3}$$

The Taylor development of the function $f(x+zh)$ is

$$f(x+zh) = \sum_{n=0}^{\infty}\left[\frac{d^n}{da^n} f(x+ah)\right]_{a=0}\frac{z^n}{n!}$$

for $|z| < \rho(x,h)$. From (3. 3. 3) we obtain

$$f(x+zh)=\sum_{n=0}^{\infty}\frac{1}{n!}\delta^n f(x,h)z^n=\sum_{n=0}^{\infty}\frac{1}{n!}\delta^n f(x,zh).$$

Hence the following result holds.

Theorem 3. 3. 4. *Let $f(x)$ be G-differentiable in the finitely open Set D. Then for $x\in D$ and $x+h$ in the C-star about x in D, we have*

$$f(x+h)=\sum_{n=0}^{\infty}\frac{1}{n!}\delta^n f(x,h). \tag{3. 3. 4}$$

Theorem 3. 3. 5. *Let $f(x)$ be G-differentiable in the finitely open set D and $\rho(x,h)>1$. If $\|f(x)\|\leqslant M$, in any C-star $C^*(a)$ about $x=a$, then*

$$\|\delta^n f(a,h)\|\leqslant Mn! \text{ for } a+h\in C^*(a). \tag{3. 3. 5}$$

Proof.
$$\delta^n f(x,h)=\left[\frac{d^n}{dz^n}f(x+zh)\right]_{z=0}=\frac{n!}{2\pi i}\int_C\frac{f(x+zh)}{z^{n+1}}dz, \tag{3. 3. 6}$$
where C is any circle $C(0,\rho')$ with $\rho'<\rho(x,h)$. In particular we may choose $\rho'=1$, so (3. 3. 5) holds.

Theorem 3. 3. 6. *Let $f(x)$ be G-differentiable in the open domain D. If $f(x)=\theta$ throghout some sphere, then $f(x)\equiv\theta$ in D.*

Proof. Suppose that $f(x)=\theta$ throughout the sphere $S_a: \|x-a\|<r$. Then (3. 3. 6) shows that $\delta^n f(a,x-a)=\theta$ for all x. It follows from Theorem 3. 3. 3 that $f(x)\equiv\theta$ in the largest sphere S with center at $x=a$ which is contained in D.

Now if b is any point in D, we may join the points a and b by piecewise continuous line segments and hence by a finite chain of open spheres in D, $S_0=S, S_1, \cdots, S_n$, such that S_k contains the center of S_{k+1}. Since $f(x)\equiv\theta$ in S_0, the preceding argument shows that $f(x)\equiv\theta$ in S_1, and hence by induction, $f(b)=\theta$. Thus $f(x)\equiv\theta$ in D. Q. E. D.

Remark. The content in this section can be extended to a locally convex space by the method in next section.

3.2. VECTOR-VALUED REGULAR FUNCTIONS IN LOCALLY CONVEX SPACES

Let E be a complete Hausdorff locally convex space on the real or complex domain K, and P be the sufficient directed set of seminorms which generates the topology of E.

Definition 3.3.6. we say that a sequence $\{x_n\}$ of elements in E is convergent strongly to $x_0 \in E$, if for any $p \in P$, $\lim\limits_{n\to\infty} p(x_n - x_0) = 0$.

Defintion 3.3.7. Let E be a complete Hausdorff locally convex space and D be a domain in the complex plane C. Suppoes $f(z)$ is a function from D to E.

(i) $f(z)$ is called a vector-valued regular function on d if for any $\varphi \in E'$, $\varphi(f(z))$ is a numerical regular function, where E' is the dual space of E.

(ii) If for any z in D and any p in P there exists $A \in E$ such that

$$\lim_{\alpha\to 0} p\left(\frac{f(z+\alpha)-f(z)}{\alpha} - A\right)$$

exists, then $f(z)$ is called strongly differentiable at z, and denoted by $f'(z)$. $f(z)$ is called strongly differentiable on D if $f(z)$ is strongly differentiable at each point z in D.

Theorem 3.3.7. *Let E be a locally convex complete linear topological space. $f(z)$ is strongly differentiable in D if and only if $f(z)$ is a vector-valued regular function.*

Proof. The necessity of the Theorem is obvious, we only prove the sufficiency of the theorem. Let S be a compact subset in D. For any $\varphi \in E'$ there exists a constant $m(\varphi, S)$ such that for $z, z+\alpha, z+\beta$ in S

$$\left|\frac{1}{\alpha-\beta}\left\{\frac{1}{\alpha}(\varphi(f(z+\alpha)-f(z)) - \frac{1}{\beta}(\varphi(f(z+\beta)-f(z))\right\}\right| \leqslant m(\varphi, S).$$

Put $f(z,\alpha,\beta) = \frac{1}{\alpha-\beta}\left\{\frac{1}{\alpha}(f(z+d)-f(z)) - \frac{1}{\beta}(f(z+\beta)-f(z))\right\}$. Then

$$J = \{f(z,\alpha,\beta) : \text{for any } z, z+\alpha, z+\beta \in S\}$$

is a weakly bounded set with elements in E. Since E is a complete locally convex space, J is a bounded set in E. Without loss of generality, we may assume U is a absorbing set, so there exists a in $\mathscr{C}$ such that $J \subset aU$, i.e.

$$\frac{1}{\alpha}(f(z+\alpha)-f(z)) - \frac{1}{\beta}(f(z+\beta)-f(z)) \in (\alpha-\beta)aU.$$

Since U is a balanced absorbing set, $(\alpha-\beta)\alpha U \subset U$ for sufficiently small α and β. By the completeness of E we get that the strong derivative $f'(z)$ exists. It follows that

$$\frac{1}{\alpha}(f(z+\alpha)-f(z))-f'(z)\in U \text{ as } \beta\to 0. \qquad \text{Q. E. D.}$$

We can extend theories about vector-valued regular functions in Banach spaces to locally convex complete spaces. For example the theorems of Sec. 3. 4 hold in locally convex complete spaces. In order to better explain these results, in the following we extend the theories of Sec. 3. 5 to complete locally convex spaces.

Theorem 3. 3. 8. *Suppose that D is a simply connected domain, that E is a complete seminorm space, and that the set $\{f(z)\}$ of vector-valued regular functions from D to E is uniformly bounded on some compact set S in D. Then $\{f\}$ is equicontinuous on S.*

Proof. For any $z_0\in D$ and $p\in P$ by the uniformly bounded condition we conclude that there exist M and r such that

$$p(f(z))\leqslant M \quad \text{as} \quad z\in S(z_0,2r). \tag{3. 3. 7}$$

From the vector-valued Cauchy integral formula we obtain

$$p(f'(z))=p\left(\frac{1}{2\pi i}\int_{C(z_0,2r)}\frac{f(t)}{(t-z)^2}dt\right), \tag{3. 3. 8}$$

where $z\in\overline{S}(z_0,r)$. (3. 3. 7) and (3. 3. 8) show that $p(f'(z))\leqslant\frac{2M}{r}$. Since S is a compact set,

$$p(f'(z))\leqslant M_1, \text{ as } z\in S,$$

where M_1 is a constant.

Applying the vector-valued Cauchy theorem we have

$$f(z_2)-f(z_1)=\int_{z_1}^{z_2} f'(z)dz \quad \text{for } z_1, z_2 \text{ in S.}$$

It follows that for any p in P

$$p(f(z_2)-f(z_1))\leqslant M_1|z_2-z_1|,$$

thus $\{f\}$ is equicontinuous on S. Q. E. D.

Theorem 3. 3. 9. *If E is a complete locally convex space, then Theorem* 3. 1. 19 *holds in E.*

Proof. Suppose that S is any compact in D and that $\{f\}$ is normal on D. For any $\varepsilon>0$, any $\{z_n\}\subset S$ and any $\{f_n\}\subset\{f\}$ by the hypothesis there exist N, $f_0(z)$ and $\{f_{n_k}\}\subset\{f_n\}$ such that

$$p(f_{n_k}(z)-f_0(z))<\varepsilon \quad \text{as } z\in s \text{ and } k>N. \tag{3.3.9}$$

It follows from a result (see Sec. 7.2 of [34], Theorem 5) that there exist positive numbers $a_1, \cdots, a_n$ and seminorm $p_1, \cdots, p_n \in P$ such that

$$|\varphi(f_{n_k}(z)-f_0(z))|\leqslant\sum_{i=1}^{n} a_i p_i(f_n(z)-f_0(z)) \tag{3.3.10}$$

z in S. so $\{\varphi(f_{n_k}(z))\}$ uniformly converge to $\varphi(f_0(z))$. Thus $\varphi(f_0(z))$ is a numerical regular function. It follows that $f_0(z)$ is a vector-valued regular function.

Without loss of generality we may assume $z_{n_k}\to z_0$ as $k\to\infty$. Hence for any $p\in P$, *where* $\{z_{n_k}\}\subset\{z_n\}\subset S$

$$p(f_0(z_{n_k})-f_0(z_0))<\frac{1}{2}\varepsilon \quad \text{as } k>N_1. \tag{3.3.11}$$

Since $\{f_{n_k}(z)\}$ is uniformly convergent, there exists N_2 such that

$$p(f_{n_k}(z)-f_0(z))<\frac{1}{2}\varepsilon \quad \text{as } z\in S \text{ and } k>N_2. \tag{3.3.12}$$

(3.3.11) and (3.3.12) give

$$p(f_{n_k}(z_{n_k})-f_0(z_0))\leqslant p(f_{n_k}(z_{n_k})-f_0(z_{n_k}))+p(f_0(z_{n_k})-f_0(z_0))$$

$$\leqslant\frac{1}{2}\varepsilon+\frac{1}{2}\varepsilon=\varepsilon.$$

Therefore $\{f_n(z_n)\}$ has a convergent subsequence, i.e. the range of $\{f\}$ in any compact set $S\subset D$ is sequentially compact.

If the range of $\{f\}$ in any compact set $S\subset D$ is sequentially compact, then $\{f\}$ is normal on D. The proof is same as that of Theorem 3.1.19. Q.E.D.

BIBLIOGRAPHY

1. Ahlfors, L. V., *Complex Analysis*, 3rd Ed., McGraw-Hill Book Co., New York, 1979.
2. Caratheodory, C., *Funktionentheorie*, Vols. I and II. Verlag Birkhäuser, Basel, 1950. Translated by Steinhardt, F., *Theory of Functions*, Chelsea Publishing Company, New York, 1954.
3. Conway, J. B., *Functions of One Complex Variable*, 2nd ed., Springer-Verlag, New York, 1978.
4. Curtain, R. F. and Pritchard, A. J., *Functional Analysis in Modern Applied Mathematics*, Vol. 132 in Mathematics in Science and Engineering, Academic Press Inc., London, 1977.
5. Diestel, J., *Sequences and Series in Banach Spaces*, Springer-Verlag, New York, 1984.
6. Diestel, J. and Uhi, J. J., *Vector Measures*, the Amer. Math. Soc., Providence, 1977.
7. Dineen, D., *Complex Analysis in Locallty Convex Spaces*, North-Holland Publishing Company Amsterdam, New York, 1981.
8. Duren, P. L., *Theory of H^p-space*, Academic Press, New York, 1970.
9. Gelfand, I., Abstrakte Funktionen und lineare Operatoren, матем сб. 4(1938), 235-284.
10. Griffel, D. H., *Applied Functional Analysis*, Ellis Horwood Limited, 1981.
11. Hille, E. and Phillips, R. S., *Functional Analysis and Semi-Groups*, the Amer. Math Soc. Colloq. Publ., Amer. Math. Soc., Providence, R1., 1957.
12. Hoffman, K., *Banach Spaces of Analytic Functions*, Prentice-Hall, Englewood Cliffs, N. J., 1962.

13. Hu, C. G., A necessary and sufficient condition of the normal family of vector-valued holomorphic functions, *Acta Sci. Natur. Univ. Nankaiensis*, 1(1981), 27-30.

14. Hu, C. G., Vector-valued integrals of Cauchy type and their applications to vector-valued singular integral equations, *Acta Sci. Natur Univ. Nankaiensis*, 1(1982), 48-58.

15. Hu, C. G., Boundary properties of a class of vector-valued analytic function in the unit disc, *Acta Sci. Natur. Univ. Nankaiensis*, 2(1984), 109-116

16. Hu, C. G., Advance of vector-valued analytic functions and its applications to vector-valued singular integral equations, *Applied Mathematics and Calculated Mathematics*, 3(1984), 95-110.

17. Hu, C. G., *Complex Analysis*, Nankai Univ. Publ. House, Tianjin, 1985.

18. Hu, C. G., *Vector-Valued Analysis and Their Applications*, Nankai Univ. Publ House, Tianjin, 1985.

19. Hu, C. G., The disturbance of vector-valued doubly-periodic Riemann boundary value problems, *J. Math. Anal. Appl.*, 131(1988), 373-391.

20. Hu, C. G., Vector-valued boundary value problems of a class, in "*Proceedings of the international Conference on Integral Equations and Boundary Value Problems*" (ed. by G. Wen and Z. Zhao), World Sci., Singapore, New Jersey, London, Hong Kong, 1991, 57-64.

21. Hu, C. G. and Beg, I., Cauchy-Stieltjes integral in locally convex spaces, *Complex Variables*, 15(1990), 233-239.

22. Li, G. P., Guo, Y. Z. and Chen, Y. T., *Automorphic Functions and Minkowski Functions*, Science Press, Peking, 1979.

23. Lindenstrauss, J. and Tzafriri, L., *Classical Banach Spaces I*, Springer-Verlag, Berlin, Heidelberg, New York, 1977.

24. Lu, J. K., Immediate methods of the finding solutions of integral equations, *Acta Sci. Natur. Univ. Wuhanensis*, 1(1975), 12-27.

25. Lu, J. K., Doubly-periodic Riemann boundary value problems, *Acta Sci. Natur. Univ. Wuhanensis*, 3(1979), 1-10.

26. Lu, J. K., *Boundary Value Problems of the Complex Analysis*, the Science and Technology Publ. House, Shanghai, 1987.

27. Markushevich, A. I., *Theory of Functions of A Complex Variable*, Chelsea Publishing Company, New York, 1985.

28. Muskhelishvili, N. I., *Singular Integral Equations*, 3rd. ed., Nauka, Moscow (in Russian), 1968.

29. Natason, N. P., *Theory of Real Variable*, The National Technological Theory Press of USSR, Moscow (in Russian), 1957.

30. Riesz, F and Nagy, B. Sz., *Functional Analysis*, *Lecons D' analyse Fonctionelle*, Translated from the 2nd French ed. by Leo F. Boron, Budapest, 1956.

31. Schaeffer, H. H., *Topological Vector Spaces*, Graduate Texts in Math 3, Springer-Verlag, New York, Heidelberg, Berlin, 1971.

32. Segal, Sanford L., *Nine introductions in Complex Analysis*, North-Holland Publishing Company-Amsterdam, New York, Oxford, 1981.

33. Wermer, J., *Banach Algebras and Several Complex Variables*, Springer-Verlag, New York, Heidelberg, Berlin, 1976.

34. Wilansky, A., *Modern Methods in Topological Vector Spaces*, McGraw-Hill, New York, 1978.

INDEX OF SYMBOLS

INDEX